迴味．

杭州菜

Hangzhou
Cuisine

陳家廚坊
Chan's Kitchen

前 言

江南好，最憶是杭州；山寺月中尋桂子，郡亭枕上看潮頭，何日更重遊？

這是唐代著名詩人白居易的《憶江南》三首中的名句。杭州之美，美在風景優美，美在深厚的歷史文化底蘊，美在難以形容的江南韻味，就像一位秀外慧中的古代才女，精緻典雅，令人難以忘懷，悠然迴味。

我們曾在上世紀八十年代、九十年代，和 2008 年三次到過杭州，不同的年代到杭州，體驗皆不同。杭州在變，我們在變，但杭州變得更快。八十年代的杭州，遍地灰藍色；九十年代的杭州，遍地建築灰塵；2008 年的杭州，遍地高樓和大酒店。三十年間，杭州市變得七彩繽紛，杭州人變得開心時麾，杭州成為了遊人如鯽的大城市。2016 年的金秋九月，桂花飄香，在杭州舉行的 G20（國際經濟合作論壇）第 11 次領導人峰會，吸引了全球的目光。中國作為東道主，為舉辦這次 G20 峰會做足準備，而杭州市是主辦城市，更力求完美，絕對是傾城之作。

杭州風味 繁迴舌尖

家母是浙江紹興人，我是吃浙江菜長大的。自從我們多次到杭州旅遊和工作後，好些年來，我都有個心願，就是撰寫一本杭州菜食譜書，但卻一直心中怯懦，猶豫不決。杭州深厚的千年歷史和飲食文化，對我而言，心中充滿敬畏。得悉 2016 年杭州舉辦 G20 峰會，我們隨即決定把杭州菜列入寫作計劃中，開始歷史文化和菜式的研究。同年底，我們有幸與一班國內和台灣的飲食界翹楚，一起食遊杭州、台州及溫州，此行獲益良多，在以西湖春天飲食集團為主的眾位好朋友的支持和幫助下，使這本書終於在 2017 年中順利出版。

本書選擇了四十多道膾炙人口的杭州名菜，更特別介紹了不少有千年歷史的南宋時期的古杭州名菜，通過我們獨特的解讀、詮釋和改良做法，使之適合普通家庭烹調，這

是本書獨特之處。地道的杭州菜，用上很多當地特有的食材，例如著名的鱘魚、千島湖的白魚、鮮活的河蝦、翠綠的西湖蒓菜、爽嫩的筍衣、春筍和杭椒、杭州的雞毛菜和地道的杭州老豆腐等。遺憾的是，一方水土一方食材，離開當地了這些材料就很難在市場上買得到，因此無法寫入這本書中；而杭州菜中有些名菜，做法比較複雜，例如傳統的杭州醬鴨和輕如無重的魚丸，考慮到讀者難以在家烹調，也只好放棄；由於醬油的品種實在太多，味道和濃淡不一，所以我們還是選用更多人認識而到處都買得到的生抽和老抽，未達盡善盡美的地方，還請讀者們理解。希望讀者能通過細讀本書，以及親自下廚，認識杭州菜，品嘗杭州菜，欣賞杭州菜。

筆錄美味　與眾分享

陳家廚坊至今已在香港出版了 13 本食譜書，幸獲廣大讀者青睞，成為暢銷的食譜書，並遠銷海外及國內市場。更令我們感到自豪和欣慰的是，2016 年我們在著名的世界食譜書系列中，代表中國菜用英文撰寫了 *China The Cookbook*，介紹中國 33 個省市自治區的飲食文化和超過 650 個各省地道菜式的食譜。這本書得到國際上很多好評，並為世界各大主要圖書館收藏，現正翻譯成法文、德文、中文、意大利文、荷蘭文、西班牙文等多國文字。2016 年秋天應出版商 Phaidon Press 的邀請，我們曾到加拿大、美國、英國及澳洲等國多個大城市進行巡迴推廣、演講及接受傳媒採訪，讓我們以香港作家的身份，有機會為中國菜在國際舞台上的發展作出貢獻，為香港人爭光。

能夠有這樣的成績，我們首先要感謝讀者們的支持，和出版社萬里機構各位同事的辛勞，您們陪伴我們一步一步地一路走來，由香港走上世界的出版舞台，衷心感謝，無限感恩！

感謝各位杭州菜飲食界朋友和專業杭菜大師的指導賜教，使本書得以順利出版。僅以此書，獻給孜孜不倦，致力把杭州菜發揚光大的各位尊敬人仕！

方曉嵐
2017 年 5 月於香港

目 錄

5

千年美食杭州菜

中華民族的食文化，歷史悠久，充分體現到文化內涵的豐富多彩，博大精深，是人民智慧的結晶，也是世界食文化的主要組成部份。中菜已經走過了四、五千年的歷史，菜餚品種之多，烹調技術之細膩，味道之複雜多樣，為世界飲食之冠。

中國是一個幅員廣大多民族的國家，由於受地理、氣候、歷史、文化傳統等影響，直接影響到當地人的生活習慣和飲食習慣，形成了各地風味獨特的菜餚，而這些菜餚經過了長時間的考驗，仍然廣受當地人的歡迎，並一代一代地承傳下去，再以省份為廣義劃分，就形成了今天中國人家傳戶曉的「菜系」。同時，在這些不同省份的菜系中，再細分為各地區不同的流派，這就是中菜博大精深之處。

浙江省東臨東海，位處長江以南，北連長江三角洲，氣候適中，雨水充足，四季分明。居住在浙江省的主要是漢族人，也有少數量的畬族、土家族、苗族、布依族和回族，現在全省人口約 5500

萬。隋唐時代稱為「淮揚」的地區，覆蓋今天的安徽、江蘇和浙江的一部份，以江都（揚州市）為經濟中心，杭州古稱「錢唐」，在隋朝得以賜名「杭州」。清康熙六年，朝廷改劃行政地區，正式有了江蘇、安徽、浙江等省名，而浙江菜成為菜系名稱，是始於清代。

長江下游地區，即華東地區，中國人稱之為江南，自古是魚米之鄉，因水陸食材豐富，膳食結構較為崇尚清淡，先民在七千多年前，就懂得種植穀物和漁獵，「飯稻羹魚」是這地區的主體膳食結構。

漢代之前，中國的菜餚，基本上只有南食和北食之分。魏晉時期，中原經歷了黃巾起義、董卓之亂，魏蜀吳的相互征戰，之後又陷入五胡亂華的混亂中，以至北方漢人大量南遷，人數達當時全國人口六份之一，使經濟和文化中心開始由黃河流域向南移至長江流域的華東地區，同時帶來了北方麥食文化的種植和加工技術，促進了南食和北食的相互交流，使江南地區的飲食習慣起了很大的變化。

浙江地區的羹膾菜餚，在古代的南食中早已遠近馳名。到了五代十國時期，中原地區仍是戰火不斷，位處江浙地區的吳越國，人民卻得到了幾十年的安定生活，使地區的經濟和文化有了良好的發展，具有本特色風味的菜餚，逐步形成，這時期

地區的醬油釀造、釀醋、釀酒、植物榨油等技術已趨成熟，而葱、薑、蒜等亦已在烹調中廣泛利用。

北宋時期，建都汴京（今開封市），後因金人滅宋，宋室南遷，史稱南宋。朝廷將杭州升格為「臨安府」，成為國家的政治文化中心，這使到杭州地區的商貿迅速發展，人口激增。南宋朝廷偏安一隅，從北方及中原遷來的皇室大族和豪門貴冑，仍然控制了國家的政權和社會財富，他們為杭州地區帶來了奢華生活的風氣，對杭州及整個地區的飲食影響深遠。生活習俗和飲食習慣的南北交融，也帶來了不少的中原和北方的烹調技術。北方食品的傳入，大大地豐富了杭州的點心小吃和菜餚品種，而本來平實清淡的杭州菜餚，從此增加了貴族的典雅品味，烹調技術趨向更細膩，用料也更多元化，使杭州菜逐漸由淳樸的農家風味，轉化成以官府菜和文人菜為重心的風格。這段南宋的歷史，是杭州菜的一個飛躍性的發展階段。

清代的康熙、雍正、乾隆三個朝代，是清代經濟發展的高峰期，康熙六次南巡和乾隆六次南巡，本意是為了穩定江南地區這個國家大糧倉和稅賦重地的統治，但江南繁華的美景和精彩的美食卻深深地吸引了他們，以至一次又一次地南巡。每次南巡，江南各地的官員和商人都傾力逢迎

在飲食上各出奇謀，勞民傷財的同時，帶動了奢華的食風，也促進了菜餚的發展。

　　江南地區水網交融，得天獨厚，物產豐富，造就了浙江人對飲食的講究，高於我國偏遠或氣候惡劣物產短缺的省份。浙江人生活富足，是歷代名人賢士、文人雅士的集中地，他們對越地飲食文化的沉澱產生了不可磨滅的影響；中國古代的十大名廚中便有五位來自這個地區，他們是太和公、劉娘子、蕭美人、王小余和家喻戶曉的董小宛，著名文學家詩人蘇東坡、陸游、袁枚、曹寅、唐伯虎等，他們既是文化人也是美食家，為我們留下了許多優美的文字來讚頌江南的美食文化。距今近千年過去了，當年南宋臨安府流行的菜餚，能流傳至今的並不多，幸好有文人雅仕為我們留下文獻，記載了一些古代菜式及其基本做法，其中有不少是值得參考學習的菜式。

　　歷史悠久的紹興菜是浙江菜的發源地，而後來居上的杭州菜現在是浙江菜系之首。杭州菜又稱杭菜，後稱杭幫菜，是江南美食中一顆燦爛的明珠。杭州屬於浙江省的杭嘉湖（杭州、嘉興、湖州）地區，地處美麗富饒的長江三角洲，氣候溫和，物產豐富。自古以來，杭州都是藝術文化之都，而杭州菜更因為歷代文人和美麗的西湖而名聞天下。杭州菜匯集了宮廷菜、官府菜、民間菜、食肆菜、船菜和寺院素菜，承傳了魏晉南北朝的北方菜，以及南宋京杭大菜，吸收了古會稽（紹興）博大精深的古吳越飲食文化，從而發展成南北風味交融的菜餚。

　　杭州菜的特色是輕油輕芡，清淡典雅，刀工細緻，重視材料新鮮，烹調方法靈活，層次分明。杭州菜常用的烹調技巧有爆炒、滑炒、生炒、清燉、嫩溜、氽、燜燒、蒸、糟、醉等，擅長做羹膾，巧用黃酒和醋，

講究色香味全。杭州菜的菜式基本上分為「湖上幫」和「城裏幫」兩個不同流派,「湖上幫」以鮮活的湖中水產如蝦、魚、甲魚、蛤蜊等為材料;「城裏幫」則以豬肉和雞鴨禽類為主要材料。

寺院素菜在杭州菜中享有很重要的位置,在特定的日子例如初一和十五要吃素食,已成為杭州人的傳統習慣。杭州自古佛教興盛,素有「東南佛國」之稱,東晉時期,印度僧人到杭州弘法,興建了靈隱寺等五個佛寺,後來南北朝梁武帝再賜田地擴建靈隱寺,杭州的佛教得到進一步發展。到了南宋,佛教進入全盛時期,作為首府臨安(杭州)的中心地區,西湖及附近群山已遍佈佛教寺廟。正如蘇東坡詩云:「三百六十寺,幽尋遂窮年」,意思是若想踏遍杭州的寺廟,需時以年計,可見杭州佛教之鼎盛。

說到杭州菜,不得不說到浙江省其他幾個地區的美食,這些地方的菜餚,或多或少,都出現在杭州的菜餚中。紹興市是古越國的首府會稽,歷史文化底蘊深厚,人傑地靈,物產富庶。紹興菜源自民間,比杭州菜的發展早了好幾百年,是浙江菜系列的搖籃和發祥地。紹興民風淳樸,菜餚鄉土風味濃,味道鹹鮮適中,以霉乾菜入饌,鹹鮮合一,是紹興菜的一大特色,被譽為紹興烹調的代表名片。明清時期,由於紹興釀酒業的高速發展,直接影響到

地區的菜餚風味,以酒入饌的烹調手法,特別是以黃酒和酒糟烹調的醉糟菜餚,更進一步蔚然成風,多種糟醉風味菜餚與杭州菜實際上已混為一體。在烹調中用料酒的技巧,後來流傳到整個中國南方各地,對包括江蘇、安徽、江西、福建、廣東等省的菜餚和烹調技術,影響深遠。另一個影響杭州菜的地方是金華市,位置在杭州市的南面,包括浙江的中部附近地區。南宋時期,金華火腿與義烏一帶出產的南肉,已經遠近馳名。杭州菜中不少菜餚就是以金華火腿和南肉為材料,味道鹹香醇厚,別具特色。

酒 蒸 雞

Chicken Steamed in Wine

　　釀酒技術在中國已有幾千年的歷史，桂花酒是江南地區的特產，據說已有超過三千年的歷史，在古代是一種供深宮貴族專享的甜酒，酒精度不高，有桂花香味。釀製方法曾長時期失傳，後來，據說是由清末代王孫溥傑獻出配方，經過不斷的現代化改良使產品重現，並以「桂花陳酒」享譽中外。

　　宋元時代，中國飲食文化有一個里程碑式的變化，取消夜禁，把封閉式里坊制，改變成開放式的街道店舖，飲食亦由沿用兩千多年的每日兩餐制，改為三餐，直接加速了餐飲業和釀酒業的發展。當時江南有一句俚語：「若要富，守定行在賣酒醋」，意思是賣酒和賣醋，都是興家致富的行業。

　　以酒入饌烹調，是南宋臨安菜餚的一大特色，南宋吳自牧的《夢粱錄》中記載的「酒蒸雞」，是來自千年古臨安的一道滋補宴客菜餚。這道帶湯汁的燉菜，用桂花酒來燉雞，經過長時間蒸燉，雞肉酥爛嫩滑，酒香撲鼻，湯汁濃稠。我們以《夢粱錄》原文所述的做法為基礎，作了一些改良，主要是拆去雞的腔骨，使之更香濃入味。宴客時推出如此精彩的古代宮庭大菜，表演割烹之道，絕對會讓所有人感到驚艷，贏得讚嘆！

●材料

嫩雞 1 隻淨重 1.4 千克	冬菇 5 朵
鹽 1 湯匙	紅蘿蔔 100 克
紹酒 1 湯匙	金華火腿 50 克
薑汁 1 湯匙	桂花陳酒 150 毫升
胡椒粉 1/2 茶匙	薑片 30 克
筍肉 200 克	葱 3 條（切段）

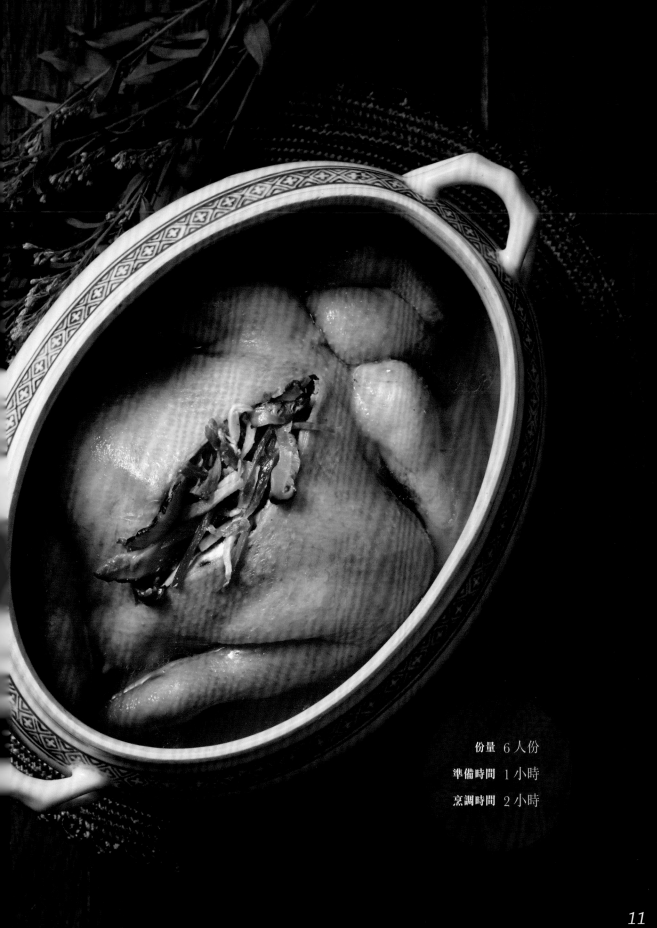

份量　6人份

準備時間　1小時

烹調時間　2小時

●做法

1. 雞洗淨，在雞腿和雞爪之間的關節下 1 厘米切斷，雞爪不要。按照下述方法把雞的腔骨去掉，用鹽、紹酒、薑汁把雞身內外擦勻，再加胡椒粉抹勻，醃半小時，備用。

2. 筍切絲，用水煮 2 分鐘瀝乾水分。冬菇用水泡軟切絲。紅蘿蔔刨絲。火腿蒸 5 分鐘後切絲。

3. 把筍絲、紅蘿蔔絲、火腿絲和冬菇絲拌勻，放入雞腔內。

4. 把雞放在大深碗或大燉盅中，倒入桂花陳酒、薑片和葱段，加蓋封密，大火蒸 2 小時。

5. 上桌後把雞身剪開，露出內餡，即成。

●烹調心得

把雞腔骨去掉，是方便在上桌進食時容易把雞切開分食。如果選擇不拆去雞胸腔骨，釀雞的餡料份量可減少三分之一。

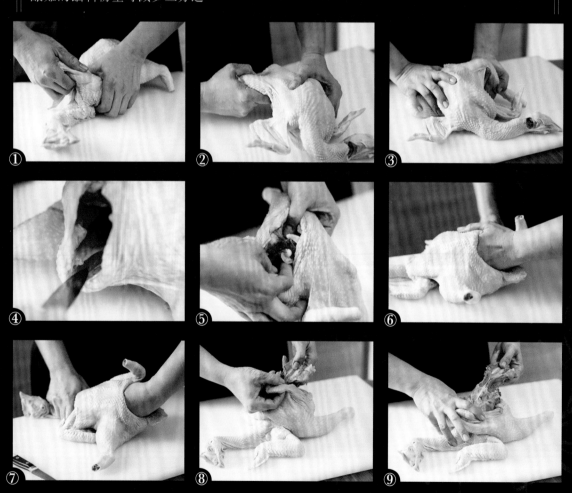

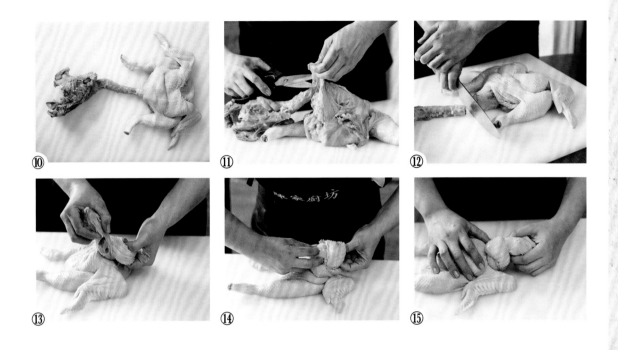

⑩ ⑪ ⑫

⑬ ⑭ ⑮

整雞拆去腔骨的方法

1. 把雞放在砧板上,胸朝下,把翅膀翻向雞背,用力拗斷翅膀和雞身之間的關節,以同樣方法拗斷雞腿與雞身的關節。①~③

2. 用一把小刀,由下端開口插進雞肉和雞胸骨之間,切開一個約 1 厘米深的小口,然後沿着雞腔把肉和骨分開。④

3. 用小刀逐漸往雞肉和雞骨之間進行切割,邊切邊把肉往雞頭方向推至約 4 厘米深。

4. 放下小刀,把手指伸進雞肉和雞骨之間,慢慢把肉和骨分離直至雞腿、翅膀和雞頸。用小刀把腿、翅膀與雞腔的關節切斷。⑤~⑨

5. 把肉和皮推向雞頭,露出整個雞腔和雞頸,再把頸和頭之間切斷,然後把雞腔骨取出。⑩~⑫

6. 把已經去掉雞腔骨的雞肉整理成雞的形狀,把雞頭和雞頸皮打成一個結,步驟完成。⑬~⑯

⑯

Chicken Steamed in Wine

Serves: 6 / Preparation time: 1 hour / Cooking time: 2 hours

Ingredients

1 chicken (1.4 kg)

1 tbsp salt

1 tbsp Shaoxing wine

1 tbsp ginger juice

1/2 tsp ground white pepper

200 g bamboo shoot

5 dried black mushrooms

100 g carrot

50 g Jinhua ham

150 ml osmanthus flavored wine

30 g ginger slices

3 scallion, sectioned

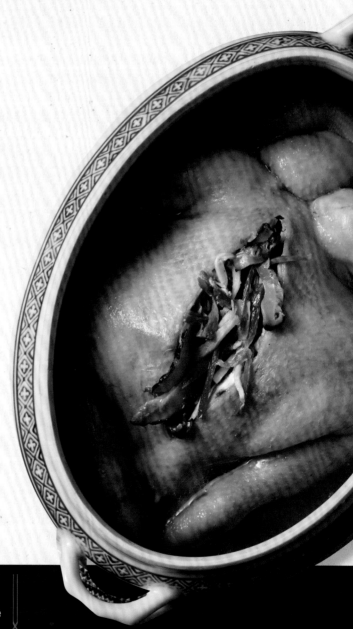

Tips:

Deboning the chicken will make serving easier. If the chicken is not deboned, the amount of stuffing can be reduced by 1/3.

Method

1. Clean chicken, and cut off feet at 1 cm below the joint between feet and leg. Debone chicken and marinate with salt, Shaoxing wine, ginger juice and ground white pepper for 30 minutes.

2. Cut bamboo shoot into strands, blanch for 2 minutes and drain. Soften mushrooms in water, remove stem, and cut into strands. Steam ham for 5 minutes and cut into thin strands.

3. Mix bamboo shoots, carrots, mushroom and ham and stuff into the chicken.

4. Place chicken in a deep bowl, put in osmanthus flavored wine, ginger slices and scallion, seal tightly and steam for 2 hours.

5. Cut open the chicken before serving in the bowl.

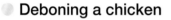

Deboning a chicken

1. Place the chicken breast side up on the cutting board. Grab the wing in one hand and the area on the body around the wing joint with the other. Pop out the wings one at a time by folding them away towards the back of the chicken body. Treat the thighs similarly to break the joints between thigh and body. ① ~ ③

2. Insert a small pointed sharp knife between skin and flesh for about 1 cm, and cut around the tail end of the chicken and the tail end of the chest bone. ④

3. DO NOT push the knife too far, just small nips. As you cut with one hand, use the other to pull the meat away from the bone to expose about 4 cm of the back bone.

4. Set the knife aside and use fingers to gently separate the flesh from the bone. Cut off the joints between the wings, thighs and the body. ⑤ ~ ⑨

5. Continue pulling out the skeleton away from the chicken flesh gently all the way up to the head. Cut off the chicken neck from the head and remove the skeleton. ⑩ ~ ⑫

6. Gently put back all skin and meat to form a chicken shape and tie a knot with the head and the skin of the chicken neck. Deboning is now completed. ⑬ ~ ⑯

份量　4-6 人份

準備時間　30 分鐘

烹調時間　45 分鐘

爐 焙 雞

Braised Chicken

　　中國古代有不少著名的女性名廚和美食家，計有五代時期的尼姑梵正，宋高宗的御廚劉娘子，南宋時期的民間名廚吳氏，明末清初的小吃名廚名妓董小宛和清代乾隆年間的譽滿江南的點心大師蕭美人。

　　南宋的吳氏，是中國歷史上最早撰寫有關飲食書籍的著名女廚師，浙江義烏人士，名字不詳。《吳氏中饋錄》是以她的姓氏命名的菜譜專錄，古代稱在家中主理廚政的婦人為「中饋」。這本古代菜譜書至為珍貴，書中分為脯鮓、製蔬、甜食三部份，記載了不少南宋時期江浙地區的民間家常菜餚技法，特別是關於醉製、醬製、臘製、醃製食物的技法，許多流傳至今，對中菜烹調技法的發展影響至深。

　　「爐焙雞」是《吳氏中饋錄》中的名菜：「用雞 1 隻，水煮八分熟，剁作小塊。鍋內放油少許燒熱，放雞在內略炒，以鏇子或蓋定，燒及熱，醋酒相拌，入鹽少許烹之，候乾再烹，如此數次，候十分酥熟取用。」「爐焙雞」是一道香味濃郁、口感酥爛的雞饌，以下做法是根據古書所載稍作修改。酒和醋是這一道菜的靈魂，太雕酒是八年以上的陳年花雕酒，顏色較深，香醇雋永。如果沒有太雕酒，可以用三年以上的紹興酒或加飯酒代替。

●材料

光雞 1/2 隻（約 700 克）　　　京葱 2 根切 5 厘米段

太雕酒／紹興酒 5 湯匙　　　　八角 1 粒

頭抽 2 湯匙　　　　　　　　　糖 1/2 湯匙

薑汁 2 湯匙　　　　　　　　　白米醋 3 湯匙

薑片 20 克　　　　　　　　　麻油 2 茶匙

●做法

1. 雞洗淨，去掉雞頭頸爪不要，斬成每塊 5 厘米大小的雞塊。加 2 湯匙酒、
   豉油、薑汁等，醃製 20 分鐘，取出雞塊，醃汁留用。

2. 鑊中燒熱約 500 毫升炸油，放入雞塊炸至金黃色，撈出雞塊，瀝油。

3. 倒出鑊中炸油，只剩 3 湯匙油，放入薑片和葱爆香，加入雞塊和醃汁，
   在鑊邊潷 3 湯匙酒，炒勻。放入八角和沸水 150 毫升，大火煮沸，轉小火，
   加蓋燜 30-40 分鐘至汁稠，取出八角、葱段及薑片，加糖和醋煮沸炒勻，
   淋上麻油，即成。

Serves: 4-6

Preparation time: 30 minutes

Cooking time: 45 minutes

Ingredients

half chicken (about 700 g)

5 tbsp Taidiao wine / Shaoxing wine

2 tbsp light soy sauce

2 tbsp ginger juice

20 g ginger slices

2 Beijing scallion in 5 cm sections

1 star anise

1/2 tbsp sugar

3 tbsp white vinegar

2 tsp sesame oil

Method

1. Clean chicken, remove head, neck and feet, and cut chicken into 5 cm pieces. Marinate 20 minutes with 2 tablespoons of wine, soy sauce and ginger juice. Take out chicken and save the marinade.

2. Heat up 500 ml of oil and deep fry chicken until golden brown. Remove chicken to drain.

3. Pour out oil leaving about 3 tablespoons in the wok, stir fry ginger slices and Beijing scallions until fragrant, add chicken and the marinade, and sprinkle 3 tablespoons of wine along the side of the wok. Stir, put in star anise and 150 ml of boiling water, and bring to a boil. Cover wok, reduce to low heat and braise for 30 to 40 minutes until sauce thickens. Discard star anise, ginger and Beijing scallions, stir in sugar and vinegar and toss, and drizzle sesame oil before serving.

糟 雞

Chicken Marinated in
Fermented Glutinous Rice

　　紹興即古代越國的首府會稽，是浙江菜系的發祥地，文化含量源遠流長，歷史比杭州菜還早了好幾百年，對後來杭州菜的發展影響很大。糟雞是紹興民間傳統名菜，是紹興酒樓食肆中的頭牌涼菜。

　　我國用糟來醃製食物，已經有二千多年歷史，北魏的賈思勰在《齊民要術》中，詳細地記載了糟的菜式，而明代李時珍在《本草綱目》中，稱糟是「藏物不敗，揉物能軟」，意思是糟製的食物不容易腐壞，而且能令食材柔軟。在沒有冰箱的年代，用鹽和酒糟來醃肉類，不失為保存食物的有效方法。

　　杭州和紹興地區的酒糟，是在用糯米釀造黃酒後剩下的酒渣，調兌後密封半年以上，即為香糟，香味濃郁，含酒精大約 8%，跟煮菜用的普通紹酒差不多。傳統上做糟雞用的是酒糟，雖然酒味特別香濃，但是因為衛生問題，酒糟基本上已經很少採用。我們的糟雞，用的是酒釀，也稱為醪糟或甜酒釀。傳統的紹興家庭喜歡自製酒釀，做法是把糯米浸透蒸熟，攤開用冷水沖至適合的溫度，加入菌種和酵母，放入罈中自然發酵而成，酒釀香味濃郁，稍帶甜味。市面上的超市及南貨店出售的盒裝或瓶裝，裏面浸着一些白色的米狀酒渣，就是酒釀。

　　醉雞與糟雞，其相同之處，兩者都是先把雞煮熟，其不同之處，醉雞是用紹興酒調汁浸漬而成的，而糟雞是用鹽和糟汁醃製的，鹽是千古不變的百味之王，糟雞的特色就是糟香入味，鹹鮮合一。有懂得美食的人認為醉與糟的分別，可以這樣來形容，「醉」，就像是欣賞一件精美的新瓷器，「糟」，卻是像欣賞一件典雅的古董。醉雞與糟雞同樣的色香味俱全，吃起來兩者都是有酒味，但細啖其韻味就是不同。

份量　4人份

準備時間　20分鐘

烹調時間　20分鐘

醃製時間　24小時

21

材料

新鮮雞 1/2 隻（約 600 克）

鹽 1.5 茶匙

酒釀 75 毫升

糖 1 茶匙

紹酒 60 毫升

做法

1. 把雞洗淨後瀝乾。

2. 燒沸一鍋水，放下雞，再煮沸後加蓋，熄火，把雞浸約 22-23 分鐘，取出放涼，切成兩大塊，再用 1/2 茶匙鹽把雞醃 30 分鐘。

3. 酒釀用攪拌機打成汁，再和 1 茶匙鹽、糖、酒及 50 毫升冷開水在大碗裏拌勻成糟汁。

4. 把雞放在大碗裏用糟汁拌勻，再連汁放在一個大食物密實袋裏，擠出空氣。

5. 把食物袋密封，放在一個大碟上，再用重物壓實，放冰箱裏 24 小時。

6. 把雞從袋中取出，斬件後淋上少許糟汁即可食用。

● 烹調心得

傳統做糟雞是放在罈子中，一層酒糟放一層紗布，再放一層雞，要密封醃七天。我們改用食物密實袋為容器，加重物壓實，這樣放冰箱中醃 24 小時已夠入味，亦不怕雞肉變壞，是一個簡單的家常好方法。

Serves: 4 / Preparation time: 20 minutes

Cooking time: 20 minutes / Marinating time: 24 hours

Ingredients

1/2 chicken (about 600 g)

1.5 tsp salt

75 ml fermented glutinous rice

1 tsp sugar

60 ml Shaoxing wine

Tips:

Using food storage bag and refrigeration in place of marinating in a traditional vase for many days is a simple method, suitable for home cooking.

Method

1. Clean and rinse chicken. Drain.

2. Bring to a boil a large pot of water, add chicken and re-boil. Cover and steep chicken in hot water for about 22 to 23 minutes, and remove chicken to cool. Cut chicken into two large pieces and marinate with 1/2 teaspoon of salt for 30 minutes.

3. Put fermented glutinous rice into a blender to blend into a sauce, and mix with 1 teaspoon of salt, sugar, wine and 50 ml drinking water into fermented glutinous rice sauce.

4. Mix chicken with fermented glutinous rice sauce in a large bowl and transfer to a large food storage bag. Seal bag after squeezing out all the air.

5. Place bag on a large plate and weigh down on the chicken with a flat heavy object. Refrigerate for 24 hours.

6. Remove chicken from the bag and cut into pieces. Drizzle fermented glutinous rice sauce on the chicken before serving.

份量 4 人份
準備時間 15 分鐘
烹調時間 30 分鐘

花雕乳鴿

Squabs in Wine

　　浙江省是魚米之鄉，這裏的先民在七千多年前，就懂得種植穀物和漁獵，自古有「飯稻羹魚」之說。杭州、紹興、嘉興、湖州等屬於平原地帶，這裏氣候溫和，水網縱橫，自古良田成片，盛產稻穀，而糯米就是這個地區的特產。

　　紹興黃酒，亦簡稱為紹酒和紹興酒，據說已有六千多年的釀造歷史。紹興黃酒產於紹興地區，以水質優良的鑒湖水和糯米釀造而成，有分土紹酒、加飯酒、善釀酒、香雪酒、元紅酒和花雕酒等分類，年份由三年至幾十年不等。紹興黃酒的酒精約為 15 度左右，口味醇厚，很適合用來烹調菜式，也可以直接飲用。據說飲紹酒能活血通絡，並含有調節營養的多酚、低肽，可以清除自由基，延緩衰老，是對身體有益的低酒精類飲品。

　　老一輩的台灣人喜歡在喝黃酒時加兩顆話梅，又或者把酒溫熱來飲，這是因為以前黃酒的釀造條件不好，特別是上世紀四、五十年代台灣釀造的黃酒，又或者是貯存條件不良時所落下的毛病，這情況猶如遇到不堪入口的葡萄酒時，有人會加入梳打水或果汁勾兌來喝一樣。不過，這種喝黃酒的習慣，卻在近幾十年由台商傳到中國，使很多人都誤以為加話梅和溫熱，才是懂得喝黃酒。其實，好品質的紹興黃酒，特別是年份高的黃酒，例如花雕和太雕，只要把酒略為冰凍，酒質更為香濃，享受已是無以尚之。

　　花雕是紹興黃酒中的上等陳年酒，酒性柔和，色澤橙黃清透，醇厚香濃，因存放在雕有花鳥花紋的密封酒罈子中，故名為花雕。在宋代時期，紹興地區家家都會以糯米釀黃酒，每當有人家生了女兒，滿月那天父母便會選幾罈好酒，用泥密封住，放在地窖中藏起來，等女兒長大出閣之時，取出陳酒以款待賓客，稱為「女兒紅」。市場上有黃酒叫做太雕，是屬於商品名稱，並不是紹興酒的傳統分類，是加飯酒和善釀酒勾對而成，一般要儲存八年以上，但顏色比較深，用來烹調菜式要注意了。

● 材料

乳鴿 2 隻　　　　　　　　花雕酒 150 毫升

鹽 1 茶匙　　　　　　　　生抽 1 湯匙

紅麴米 1 湯匙　　　　　　老抽 1 湯匙

薑片 30 克　　　　　　　蠔油 1 湯匙

蒜頭 4 瓣（切片）　　　　冰糖 30 克

乾葱 4 粒（切片）　　　　清雞湯 150 毫升

● 做法

1. 乳鴿洗淨，加鹽醃 15 分鐘，瀝乾。

2. 紅麴米用香料袋或茶袋裝好，備用。

3. 在鑊中燒 250 毫升油至中溫（約 150℃），把乳鴿放在漏勺上，用油淋至微紅色。

4. 倒出油，只留 2 湯匙在鑊中。爆香薑、蒜片和乾葱片，加酒、生抽、老抽、蠔油、冰糖、雞湯和紅麴米，煮沸。

5. 放下乳鴿，再煮沸，轉慢火煮 20-25 分鐘至乳鴿完全熟透，取出斬件放盤。

6. 大火收汁，淋汁乳鴿上即成。

● 烹調心得

1. 此道菜適宜用花雕酒，或三或五年釀的紹興酒。

2. 紅麴米是天然的染色材料，在可令乳鴿的顏色紅潤。在中藥店或食物雜貨店有售。

Serves: 4 / Preparation time: 15 minutes / Cooking time: 30 minutes

Ingredients

2 squabs

1 tsp salt

1 tbsp red yeast rice

30 g ginger, sliced

4 cloves garlic, sliced

4 shallot, sliced

150 ml huadiao wine

1 tbsp light soy sauce

1 tbsp dark soy sauce

1 tbsp oyster sauce

30 g rock sugar

150 ml chicken broth

Method

1. Clean, and marinate squabs with salt for 15 minutes.

2. Put red yeast rice in a spice bag.

3. Heat 250 ml of oil in a wok to a medium temperature (about 150°C) , hold squabs in a hand held colander over the wok and drizzle heated oil over them until skin turns slight brown.

4. Pour out oil leaving 2 tablespoons in the wok, stir fry ginger, garlic and shallot until aromatic, add wine, soy sauces, oyster sauce, rock sugar, chicken broth and red yeast rice, and bring to a boil.

5. Put in squabs, re-boil, change to low heat and simmer for 20 to 25 minutes until thoroughly cooked. Take out squabs, cut into pieces and transfer to plate.

6. Reduce the sauce and drizzle over the squabs.

Tips:

1. Shaoxing wine that has been aged 3 to 5 years can be used in place of huadiao wine.

2. Red yeast rice is a natural food coloring ingredient and can be found in Chinese medicine stores or in dry good stores in the wet market.

火膧神仙鴨

Braised Duck with Jinhua Ham Shank End

金華火腿是浙江名產，醃製金華火腿，用的是豬隻的後腿，用豬前腿醃製的是風腿。火腿分成五個部份，火爪、火膧、上方、中方、油頭。這五個部份以上方為最佳，中方次之，油頭位置因油味重，淪為再次的位置，一般用於切絲。火爪肥瞜少，最適合用來吊湯。火膧肉少，脂肪層薄，皮質多而口感甘糯。亦有寫作「火蹱」，膧和蹱兩個字都通用，都是指粵語中所謂「上五寸、下五寸」的位置。

杭州人稱火膧為「膧兒」，稱「兒」的杭州話很多，是宋室南遷帶來的北方土語。「火膧神仙鴨」是杭州傳統名菜，就是把鴨和火膧一起以微火慢燉而成，火膧肉艷紅柔軟，湯汁香濃，鴨肉酥爛油潤。江浙民間素有「爛煮老雄鴨，功效比參芪」的說法，這道「火膧神仙鴨」，被認為是營養豐富、滋陰健脾的江南美食。

這道菜的名字來自一個傳說，話說很久之前，廚師們做火膧筍乾鴨湯，因為當時沒有鐘表來計時，火候難以掌握，經常被大師傅責備。有一天有位廚師又把鴨子燉過爛了，感到十分沮喪，這時來了一個老者，教導他以後在燉鴨時，先準備三柱香，燒完一柱再燒一柱，當三柱香燒完時，鴨子也燉得剛剛好了。從此，這個法子很快就在民間傳開，人們都認為那位老者是神仙下凡打救世人，於是這道菜從此就稱為「火膧神仙鴨」。

「火膧神仙鴨」主要是吃火膧和鴨肉，而甘香的火膧更是主角，這並非一道火腿煲鴨湯，不然任何火腿的位置都可以用來煮湯。如果用整個火膧來煮鴨，煮到火膧夠酥軟，整個菜一定會太鹹，而鴨肉也會太爛，湯也會很深色。所以我們採取了分開及分段來煮火膧和鴨，煮出來的「火膧神仙鴨」鹹淡適中，湯汁鮮美，火膧口感甘腴無比。

●材料

火膧 1 隻

鹽 2 茶匙

筍乾 120 克

米鴨 1 隻（約 2.4 千克）

葱結 4 束

薑片 20 克

紹興酒 2 湯匙

小棠菜 200 克

火膧

●烹調心得

1. 買火膧一般要買整段連骨火膧，約重 1 公斤，下述做法 4 削去大骨的一段後，剩下約 400 克肉已足夠煮 1 隻鴨，其他火膧肉及骨可留他用。

2. 火膧的皮是最美味的部份，連肉一起切厚片。

3. 由於火膧有鹹味，要試味後再決定是否需要加鹽。

做法

1. 沿火䐹大骨切成兩邊,帶大骨的一邊留起作他用。如果剩下的一邊帶有小片骨頭,要先把骨頭剔除。

2. 火䐹洗淨,放入一大盆水中,水要完全浸過火䐹,加入鹽拌溶,把火䐹浸泡 6 小時,取出,用清水沖洗,再燒沸大鍋水氽燙 3 分鐘,取出,用清水沖洗一下,

3. 用清水洗去筍乾表面鹽分,再放入約 40℃暖水中,用手輕輕搓揉約 5 至 10 分鐘,中途可換水一次,至筍乾稍為變軟,撈出,用水沖洗,再瀝乾水分。把筍乾較硬的節位部份切去不要,再斜切成約 5 厘米長條。

4. 把火䐹放在鍋中,加清水浸過,煮沸後改小火煮約 1.5 小時,取出火䐹,湯留作後用。

5. 同時把鴨內外洗淨,切去鴨尾及多餘脂肪不要。燒沸大鍋水,放入鴨氽燙 5 分鐘,取出瀝乾。

6. 用一隻大砂鍋或鐵鍋,放入鴨,鴨胸向上,加入筍乾、葱結、薑片,加清水至浸過鴨身,加蓋。大火燒沸,改用小火煮約 2 小時至鴨肉酥爛,取出葱結和薑片不要。

7. 小棠菜洗淨灼熟,放入鴨湯中。

8. 火䐹切片,放在鴨上,倒入部份煮火䐹的湯,試味,沸煮 5 分鐘,加入紹興酒,原鍋上桌,即成。

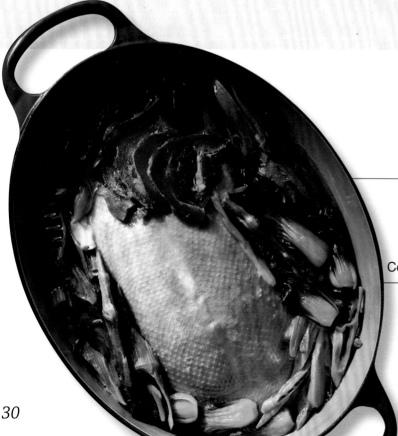

Serves: 6-8

Soaking time: 6 hours

Preparation time: 20 minutes

Cooking time: 2 hours 15 minutes

Ingredients

1 Jinhua ham shank end

2 tsp salt

120 g dried bamboo shoot

1 duck (about 2.4 kg)

4 scallion, knotted

20 g ginger slice

2 tbsp Shaoxing wine

200 g Shanghai brassica

Method

1. Cut ham shank end vertically along the large bone into two pieces and save the piece with the large bone for other uses. Remove the small piece of bone (if present) in the remaining piece used with this recipe.

2. Rinse the ham, and submerge it in a large pot of water with salt added and soak for about 6 hours. Rinse, and blanch for about 3 minutes. Rinse again.

3. Rinse dried bamboo shoots and soak in warm water (about 40°C). Rub gently for 5 to 10 minutes until soft. Change warm water once. Rinse with fresh water and cut away the knots that remain very firm. Slant cut the remainder into 5 cm slices.

4. Place ham shank in a pot covered with water, bring to a boil, reduce to low heat and simmer for 1.5 hours. Take out ham and save the ham juice for later use.

5. In the meantime, clean the duck and remove tail and excess fat, blanch duck for 5 minutes and drain.

6. Place duck in a large pot or casserole with the breast facing up, put in bamboo shoots, scallion and ginger, add water to completely cover duck and bring to a boil. Reduce to low heat, cover, and simmer for about 2 hours. Discard scallion and ginger.

7. Blanch Shanghai brassica and put in with the duck.

8. Cut ham into slices, place on top of duck, add an appropriate quantity of ham juice to flavor, and bring to a boil for 5 minutes. Add wine before serving.

Tips:

1. A Jinhua ham shank end weighs about 1 kg. After cutting away the large bone portion and save for other uses, the remaining portion which weighs about 400 g is enough for cooking with a whole duck.

2. The skin of the ham shank end is perhaps the best part and should be, sliced together with the meat.

3. The juice from cooking the ham shank end is rather salty. Salt should only be added if necessary.

杭幫菜

火臟神仙鴨

鱉蒸羊

Steamed Softshell Turtle with Mutton

魚加羊是「鮮」字，那麼，甲魚加羊肉又是一道怎樣的鮮美菜式呢？

這道菜已有千年的歷史，在南宋吳自牧《夢梁錄》中有記載。鱉即甲魚 / 水魚，「鱉蒸羊」就是用羊肉湯來蒸甲魚，是一道秋冬滋補的家庭菜式，肉質酥爛湯汁濃，其鮮味無比，令人食指大動。

甲魚，廣東人稱為水魚，古稱為鱉、黿、元魚。甲魚生長在河流湖泊中，以小魚小蝦為食，其分布很廣，長江和珠江流域，雲南、貴州、廣西、海南島都有野生甲魚。甲魚有一近親是山瑞，生長在山間的溪澗之中，樣子與甲魚甚為相似，但山瑞的頸部及背部有粗大疣粒，腹部呈黃白色，裙邊比甲魚厚和寬，價格也比甲魚貴得多。

甲魚味道鮮美，營養豐富，中醫認為吃甲魚能補中益氣，治風濕痹痛，滋補有益。吃甲魚最佳的部份是四周下垂的甲魚裙（水魚裙），味道鮮美，肉軟而甘腴，甚至連僧人也為之垂涎，見《五代史補》中曰：「南唐僧人謙光嗜鱉，國主戒之。對曰：老僧無他願，但得鵝生四腿（掌），鱉長兩群（裙）足矣，國主大笑。」意思是但願一隻鵝生四隻鵝掌，甲魚能有兩層甲魚裙，可見兩者之美味。

以前吃甲魚，是貴價的菜式，二十多年前國內發明了利用溫泉水使甲魚避開冬眠，而令其加快生長，逐漸培養成養殖甲魚的技術，甲魚的市場價格下降，使甲魚菜式得以走入平常百姓家。

份量 6-8 人份

準備時間 20 分鐘

烹調時間 2 小時 30 分鐘

33

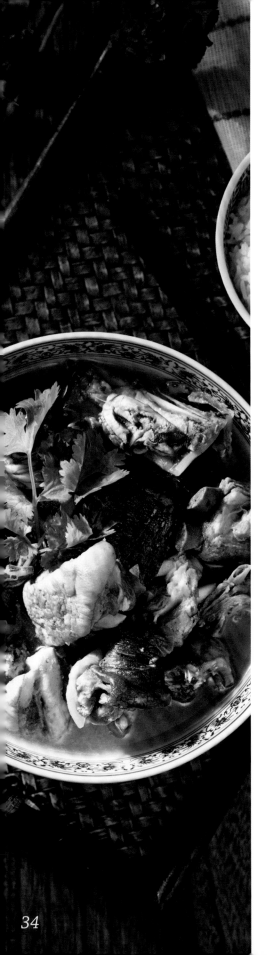

材料

羊肉 300 克	薑片 30 克
原粒白胡椒 1 茶匙	甲魚 / 水魚 1 隻（約重 500 克）
薑 30 克	葱 3 條
草果 2 粒	紹酒 3 湯匙
鹽 1 茶匙	蠔油 1 湯匙

做法

1. 羊肉切 2 厘米方塊洗淨，用滾水汆 5 分鐘後撈出，瀝水備用。

2. 胡椒輕輕拍裂，放入小香料袋 / 茶葉袋束好，薑切塊拍碎，與羊肉、草果一起放入鍋中，加 1 茶匙鹽和水 1.5 公升，大火煮沸，加蓋，轉小火慢煮約 1 小時 30 分鐘至腍。

3. 取出羊肉，大火收汁至約剩 500 毫升羊肉湯。

4. 甲魚宰淨切件，用清水泡 15 分鐘，用大沸水汆燙 3 分鐘，撈出用清水沖淨。和薑片一起放在深碗中，葱洗淨打成葱結放在上面。把煮過的羊肉塊放在葱和甲魚的上面。

5. 羊肉湯中取走薑塊、草果和香料袋，加蠔油和紹酒，注入蒸碗中，用鋁箔紙把蒸碗封密，放入蒸爐中，大火蒸 15 分鐘，即成。

●烹調心得

1. 羊肉和甲魚的烹調時間不同，在蒸的過程中羊肉不容易變腍，所以要先把羊肉用水煮至腍，才可以和甲魚同蒸。

2. 草果是一種烹調羊肉時去羶味的最佳香料。

Serves: 6-8 / **Preparation time:** 20 minutes

Cooking time: 2 hours 30 minutes

Ingredients

300 g mutton

1 tsp white peppercorn

30 g ginger

2 amomum tsao-ko

1 tsp salt

30 g ginger slices

1 softshell turtle (about 500 g)

3 scallion

3 tbsp Shaoxing wine

1 tbsp oyster sauce

Method

1. Cut mutton into 2 cm squares, blanch for 5 minutes and drain.

2. Crush white peppercorn lightly and put into a spice bag. Smash ginger and place into a pot together with mutton and amomum tsao-ko. Add salt and 1.5 liters of water, bring to a boil, cover, reduce to low heat and simmer for 1 hour and 30 minutes until tender.

3. Remove mutton, and reduce soup over high heat to about 500 ml.

4. Cut softshell turtle into pieces, clean, and soak in fresh water for 15 minutes. Drain. Blanch for 3 minutes, drain and rinse with fresh water. Place softshell turtle in a large bowl together with ginger slices and scallion, and top with mutton.

5. Remove ginger and scallion from the mutton soup, stir in oyster sauce and wine, and put into the bowl. Seal tightly and steam over high heat for 15 minutes.

Tips:

1. Mutton requires a much longer cooking time than softshell turtle and has to be cooked until soft before adding to the turtle.

2. Amomum tsao-ko helps to remove the gamy taste of the mutton.

乾菜東坡肘子

Braised Pork Knuckle with Shaoxing Mustard Greens

　　紹興有兩種特產，令紹興人感到非常自豪，日常飲食上又離不開的，就是霉乾菜和紹興酒。霉乾菜又名乾菜或烏乾菜，紹興醃製霉乾菜的技術已有過千年歷史，在清代時期曾是朝廷貢品。紹興市土地肥沃富饒，盛產品質優良的芥菜，而紹興的氣溫、潮濕，正好為生產霉乾菜提供了絕佳的天然條件。由紹興至寧波，整個地區的農村，幾乎家家戶戶都精於醃漬菜蔬植物包括各種乾菜和筍乾，其中以紹興生產的品質最佳。杭州菜和上海菜都經常用上紹興霉乾菜，紹興菜中的「霉乾菜燜肉」，被稱為紹興第一菜。

　　宋朝時中原人南遷，始經地是華東的江浙（古淮揚地區），部份人在會稽（今天的紹興市）學會了傳統醃霉乾菜的醃製技術。由於客家人甚信好意頭，「霉」字和「乾」字都不吉利，而嶺南山區醃製冬芥菜的時間，正值梅花盛開的季節，於是就有了惠州客家梅菜，但紹興乾菜的「霉」和「乾」兩字一直源用至今。紹興霉乾菜顏色較深，樣子與惠州客家梅菜相似，但由於製作中有發酵過程，所以紹興霉乾菜的香味比惠州梅菜更濃烈。

　　乾菜東坡肘子，基本上是東坡肉做法再加上紹興霉乾菜，酥爛滑糯的口感中，稍稍帶着乾菜的鹹香，再加上肘子留骨，為傳統的東坡肉增加了野性的震撼，鹹甜之間，醇厚甘腴的味蕾感受，令人垂涎三尺。

霉乾菜

材料

紹興霉乾菜 60 克	葱 4 條（切段）
紅糖 5 湯匙	紹酒 250 毫升
紅麴米 1 湯匙	生抽 3 湯匙
帶骨肘子 1 隻	冰糖 20 克
薑片 30 克	小棠菜 200 克

份量　4-6 人份
準備時間　15 分鐘
烹調時間　3 小時

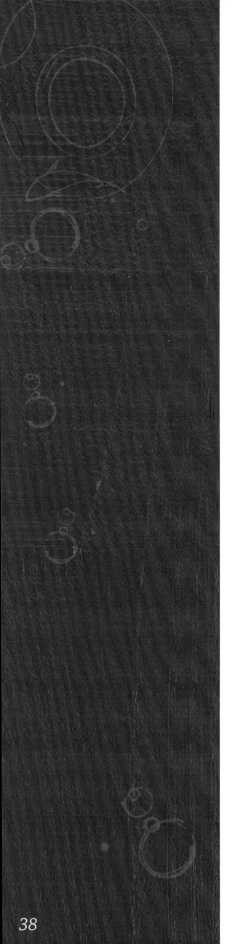

做法

1. 霉乾菜用清水洗淨後，切碎，再用 3 湯匙糖拌勻，蒸 1 小時，備用。

2. 紅麴米用香料袋袋起，備用。

3. 把肘子部份瘦肉切除，留作其他用途。刮淨肘子的皮，洗淨，再汆水 5 分鐘，取出。

4. 在小鍋裏燒熱約 2 湯匙水，放入 2 湯匙紅糖，慢火煮成焦糖，熄火，放入肘子上色，取出。

5. 在大煲裏燒熱 60 毫升油，爆香薑葱，放入霉乾菜炒香，加紹酒和紅麴米，再放入肘子，加水至覆蓋肘子，煮沸後轉小火，加蓋燜 1 小時，取出紅麴米，放入生抽和冰糖，再燜 15 分鐘。

6. 把肘子翻過來再燜約 45 分鐘至軟腍，取出裝盤。把汁收至濃稠，淋在肘子上。

7. 小棠菜另用水焯熟，排在肘子邊上，即成。

● 烹調心得

1. 肘子，是豬前腿的小腿部份，肥肉比元蹄少，肘子的皮和腳筋是主要的美味部份。香港市場賣的豬手其實就是連豬蹄的肘子，買的時候請檔主把豬蹄部份切除，斬件留作他用。

2. 豬肉經過水煮，豬皮毛孔會稍為張開，趁熱上糖色更能附上，燜煮後就有較均勻紅潤的色澤。焦糖不要煮得太濃稠，火候要控制得好，小心煮焦黏鍋。

3. 紅麴米的作用是要使肘子顏色更紅亮，賣相更好看。

Serves: 4-6 / **Preparation time:** 15 minutes / **Cooking time:** 3 hours

Ingredients

60 g Shaoxing mustard greens

5 tbsp red sugar

1 tbsp red yeast rice

1 pork knuckle with bone

30 g ginger slices

4 scallion, sectioned

250 ml Shaoxing wine

3 tbsp light soy sauce

20 g rock sugar

200 g Shanghai brassica

Method

1. Rinse mustard greens, chop, mix with 3 tablespoons of sugar, and steam for 1 hour.
2. Put red yeast rice in a spice bag.
3. Remove part of the lean meat from the pork knuckle and save for other use. Scrape skin clean of hair, rinse, and blanch for 5 minutes.
4. Heat 2 tablespoons of water in a pan, add 2 tablespoons of sugar, and caramelize over low heat. Roll pork knuckle in the pan to color skin with caramel.
5. Heat 60 ml of oil in a large pot and stir fry ginger and scallion until pungent, stir in mustard greens, add wine and red yeast rice, and put in pork knuckle. Add water to cover knuckle completely and bring to a boil. Reduce to low heat, cover, and braise for 1 hour. Remove red yeast rice, add soy sauce and rock sugar, cover and braise for another 15 minutes.
6. Flip over knuckle, braise for another 45 minutes until tender and transfer to a plate. Reduce sauce in the pot and drizzle over knuckle.
7. Blanch Shanghai brassica and put on the side of the knuckle.

Tips:

1. Pork knuckle is part of the pig foreleg which contains less fat than the hind leg. The feet can be saved for other uses.
2. Blanching the pork cleans the pork as well as making it easier to take on the color of the caramel. Take care not to overcook the caramel as it burns easily.
3. Red yeast rice gives the pork a good red color making it more appetizing.

土雞蛋百葉結燒肉

Stewed Pork Belly with Knotted Baiye and Ranch Eggs

份量 3-4 人份

準備時間 10 分鐘

烹調時間 1 小時 15 分鐘

　　廣東菜的「燒」是指燒烤，例如叉燒，是古時最原始的烹調方法「炙」，而外省菜中的「燒」，即廣東菜的「炆」，是沿用中原的古稱法，並非廣東方言，康熙字典有「炆」字，但無「燜」字。宋朝末年開始，大量中原人南遷成為了廣東人或客家人，他們對於中原文化有一定的堅持，世代流傳。這種用文火把材料煮至入味完成的水烹法烹調技巧，在北方一般都稱為「燒」，據說是因為明太祖朱元璋的長孫惠帝朱允炆，只當了四年皇帝，就遭叔叔朱棣奪權下了台，還在與政變有關的宮庭大火中被燒死，北方和江南一帶的人，為了避諱就不再用「炆」字，所以有「南炆北燒」的說法。

　　用黃豆製成的食品種類很多，比較熟悉的便是板豆腐、嫩豆腐、山水豆腐、腐竹、腐皮、豆腐泡、豆腐乾等，而百頁（又名千張）是江浙地區很受歡迎的豆製品。做法是將泡軟的黃豆加水磨成豆漿煮沸濾渣後，加凝固劑凝成「豆腐腦」，用布摺疊壓製成薄片狀豆皮。百頁結的口感比豆腐富彈性，不易破碎，可與其他食材直接烹煮，最適合用來煮湯或滷煮。用百頁做的江南菜餚有：醃篤鮮、雪菜毛豆炒百頁、土雞蛋百頁結燒肉。

　　農家放養的「走地雞」，是吃蟲子長大的，生的蛋叫做土雞蛋，山上放養的雞所生的蛋，在北方也叫柴雞蛋。放養的雞不餵飼料，所以土雞蛋體形較一般雞蛋小，現在超市中有各種標榜原生態的農家土雞蛋出售，事實上有多「土」就不知道了。

材料

土雞蛋 3 隻　　　　　　　五花腩 300 克

百頁結 100 克　　　　　　老抽 2 湯匙

薑 20 克　　　　　　　　　蠔油 1 湯匙

蒜頭 4 瓣　　　　　　　　糖 1/2 湯匙

做法

1. 雞蛋冷水下鍋，煮沸後轉小火再煮約 5 分鐘，涼卻後去殼備用。

2. 百頁結洗淨，瀝乾水分，備用。

3. 薑切粒，蒜頭去衣切半。

4. 五花腩汆水，瀝乾，再切成約 2 厘米件。

5. 在鍋中燒熱 2 湯匙油，爆香薑蒜，放入五花腩大火炒約 2 分鐘，讚酒，放入老抽、蠔油和糖，加約 500 毫升水，煮沸後轉小火，加蓋煮約 60 分鐘或至腍。

6. 放入百頁結和雞蛋煮 5 分鐘，大火收汁即成。

●烹調心得

1. 冰鮮的百頁結可以在南貨舖買到。

2. 建議購買做好的冰鮮百頁結，若買的是黃色的乾百頁皮，便要先經過處理才能做菜。用小蘇打（baking soda 或 bicarbonate soda，亦稱食粉）開水浸百頁，可令乾百頁軟化及顏色變白。方法是把 1 茶匙蘇打粉放進 1 公升溫水裏拌溶，把乾百頁放進，浸泡約 15 分鐘到顏色變白，再用清水徹底漂洗百頁，洗去蘇打味，然後打結。

Serves: 3-4 / **Preparation time**: 10 minutes / **Cooking time**: 1 hour 15 minutes

Ingredients

3 ranch eggs

100 g knotted baiye

20 g ginger

4 cloves garlic

300 g pork belly

2 tbsp dark soy sauce

1 tbsp oyster sauce

1/2 tbsp sugar

Method

1. Place eggs in a pot of cold water, bring to a boil, reduce to low heat and cook for about 5 minutes. Shell eggs after cooled.

2. Rinse baiye and drain.

3. Chop ginger, and peel and cut each clove of garlic in 2 halves.

4. Blanch pork belly, drain, and cut into 2 cm pieces.

5. Heat 2 tablespoons of oil in a casserole, stir fry ginger and garlic until aromatic, add pork belly and stir fry over high heat for about 2 minutes. Sprinkle wine, add soy sauce, oyster sauce, sugar and about 500 ml water, bring to a boil, cover, reduce to low heat and simmer for about 60 minutes or until tender.

6. Put in baiye and eggs and cook for 5 minutes. Reduce the sauce over high heat before serving.

Tips:

1. Chilled knotted baiye can be found in shops selling Shanghainese foods.

2. If dried baiye sheets are used, soak in 1000 ml warm water for about 15 minutes with 1 teaspoon of baking soda added until they turn into a milky color. Rinse thoroughly with fresh water to rid of the taste of baking soda before knotting.

份量　4-6 人份
準備時間　15 分鐘
烹調時間　10 分鐘

火腿蠶豆

Stir Fried Broad Beans with Ham

　　蠶豆，原產於中亞及東非地區，在西漢時期傳入中國，所以又名胡豆、羅漢豆。據李時珍的《本草綱目》記載：「豆莢狀如老蠶，故名。」蠶豆有一種特殊的香味，可作為時令蔬菜，亦可製作小食、豆汁和豆泥，也有用來製醬，四川菜常用的豆瓣醬，就是以蠶豆作為主要材料。

　　每年春夏之交，是江南地區蠶豆上市的季節，據說南宋著名的詩人楊萬里，很喜歡吃鮮嫩的蠶豆，作詩曰：「翠莢中排淺碧珠，甘欺崖蜜軟欺酥」，來形容蠶豆的淡綠酥軟。當時新鮮蠶豆的吃法一般是連皮煮熟，剝皮吃豆肉，有一年到了蠶豆上市的季節，剛好楊府的老廚師不在，新來的廚師知道主人愛吃蠶豆，就別出心裁地做一味炒蠶豆的菜式。他用金華火腿切丁來炒蠶豆，楊萬里吃後大為稱讚。從此，這味清香酥軟、美觀悅目的「火腿蠶豆」，就成了楊府的名菜，後來此菜更流傳開來，成為一道經典的杭州菜。

　　這味「火腿蠶豆」，做法很簡單，在蠶豆上市的季節，菜市場上有連豆莢的新鮮蠶豆出售，撕開豆莢取出淡綠色的蠶豆肉，再剝去豆眉和豆皮便可烹調；近年市場上有袋裝的冰鮮去莢蠶豆肉出售，更為方便。

●**材料**

蠶豆肉 300 克　　　　　　糖 1 茶匙

鹽 1/4 茶匙　　　　　　　清雞湯 2 湯匙

金華火腿 75 克　　　　　　麻油 1/2 茶匙

●**做法**

1. 剝掉蠶豆肉的豆眉和外層的豆皮，洗淨，放入沸水中焯 3 分鐘，撈出備用。

2. 火腿隔水蒸 3 分鐘，待涼，切成 0.3 厘米丁方的火腿粒。

3. 在鑊中燒熱 2 湯匙油，把蠶豆肉快炒約 1/2 分鐘，加入鹽、火腿粒、糖和清雞湯，炒至收汁，加麻油炒勻，即成。

Serves: 4-6

Preparation time: 15 minutes

Cooking time: 10 minutes

Ingredients

300 g podded broad beans

1/4 tsp salt

75 g Jinhua ham

1 tbsp sugar

2 tbsp chicken broth

1/2 tsp sesame oil

Method

1. Slit the thin skin that covers each bean and push the bean out. Rinse and blanch for 3 minutes. Drain.

2. Steam ham for 3 minutes and dice into 0.3 cm cubes.

3. Heat 2 tablespoons of oil in a wok, stir fry beans for about 1/2 minute, then add salt, ham, sugar and chicken broth, and stir fry until thicken. Stir in sesame oil before serving.

東坡肉

Dongpo Pork

蘇東坡是北宋最著名的文人，無論在詩、詞、賦、散文、書、畫都有很高的成就，除了在文學上的修養，他更是著名的美食家。蘇東坡生性耿直，得罪的人不少，仕途坎坷；但他為官清正，在做地方官的時候，深得人民愛戴，也因此留下很多關於他的故事。

東坡肉有幾個不同版本的典故，其中一個說的是蘇東坡有次路過贛北永修，因為當地農民誤解了蘇東坡的說話，做出了一個用稻草和豬肉同煮的紅燒肉。紅燒肉是否會因為有稻草而變得更好吃，不得而知；但是現在很多菜館做的紅燒肉，上面繫一根稻草，叫做「東坡稻草肉」，據說就是源於這個故事，當然，繫上稻草，也不失為把燉煮中的豬肉翻轉的好辦法。

還有一個東坡肉的故事，是發生在湖北省的黃州，當時蘇東坡被貶為黃州協團練副使。被貶後的薪俸不多，生活困苦，請得廢地數十畝，自耕自足，建了一間草堂，自號東坡居士，閑來研究烹飪技術。當時黃州的豬肉價賤，蘇東坡最喜歡吃豬肉，他寫了一首《豬肉頌》總結他烹調豬肉的經驗：「淨洗鐺，少著水，柴頭竈煙焰不起。待他自熟莫催他，火候足時他自美。黃州好豬肉，價賤如泥土。貴者不肯食，貧者不解煮。早晨起來打兩碗，飽得自家君莫管」。只有像蘇東坡這樣天才橫溢、安貧樂道的人，才能在困境中還有心情和耐性，把廉價的食材研究出美食來。蘇東坡後來當上了杭州知府，東坡肉更被廣泛流傳，成為杭州名菜。

份量 4-6 人份
準備時間 10 分鐘
烹調時間 3 小時

●材料

五花腩肉 600 克
紅糖 2 湯匙
紅麴米 1 湯匙
薑片 40 克
紹興酒 400 毫升
冰糖 20 克
生抽 1.5 湯匙
老抽 1.5 湯匙

●烹調心得

1. 把豬肉用清水煮有幾個作用，一是氽去生豬肉的血水和污垢，二是豬肉經過水煮，豬皮毛孔會稍為張開，這時趁熱上糖色，便更能附在豬皮上，燜煮後就會有均勻紅潤的色澤。三是煮完的五花腩比較容易切得整齊。

2. 焦糖不需要煮得太過濃稠，火候要控制得好，小心煮焦沾鍋。

3. 紅麴米的作用是要東坡肉顏色更紅亮，賣相更好看。

4. 請賣肉店家不要用火槍去燒豬皮上的毛，因為豬皮經過火燒會縮小，整塊肉變得肉多皮少，煮出來的東坡肉就不好看了。

做法

1. 五花腩肉整塊刮去細毛洗乾淨，豬皮向下放在砧板上，用刀把最上面一層的全瘦肉切去不要。切出來的瘦肉可以留作他用。

2. 把五花腩肉放進一鍋沸水中大火煮肉 20 分鐘後，用清水沖 5 分鐘，瀝乾後，用刀把瘦肉部份切成四塊正方型，但豬皮不切斷。

3. 把紅糖放入鍋，開小火用 2 湯匙水慢煮紅糖，不斷用勺攪拌，煮成焦糖，然後把五花腩肉皮朝下放在焦糖上沾上糖色，再把豬肉反過來，讓焦糖沾滿所有位置，熄火取出五花腩肉。

4. 把紅麴米用 250 毫升水慢火煮 15 分鐘，把渣過濾後，紅麴米水留用。

5. 用一隻小鍋，把薑片鋪在鍋底，五花腩肉皮朝下放在薑片上，加進紹酒、紅麴米水和清水至浸過材料，加蓋用大火煮沸，改用小火煮半小時後，放入冰糖、生抽和老抽煮 1 小時，然後把五花腩肉反轉使豬皮朝上，再煮 1 小時，熄火取出裝盤。把鍋底的稠汁淋在豬皮上，即成。

Serves: 4-6 / Preparation time: 10 minutes / Cooking time: 3 hours

Ingredients

600 g pork belly

2 tbsp red sugar

1 tbsp red yeast rice

40 g ginger slices

400 ml Shaoxing wine

20 g rock sugar

1.5 tbsp light soy sauce

1.5 tbsp dark soy sauce

Method

1. Scrape pork skin clean of bristles, remove the top layer of lean meat and save for other uses.

2. Blanch the pork for 20 minutes and rinse with cold water for 5 minutes. Drain. Cut the lean meat into four equal quarters but without cutting through the skin.

3. Caramelize red sugar with 2 tablespoons of water, place the pork belly skin down first and then turn over to ensure the pork belly surface is fully covered with caramel.

4. Boil red yeast rice in 1 cup of water for 15 minutes, strain, and keep the red yeast rice water for later use.

5. Line the bottom of a pot with ginger slices, on top of which put the pork belly with skin facing down. Add wine, red yeast rice water and cold water until the pork belly is totally covered. Bring to a boil and reduce to low heat to simmer for 30 minutes, then add rock sugar, soy sauces and simmer for 1 hour. Turn over the pork belly and simmer for 1 more hour. Remove pork belly to a plate with skin side up and drizzle the juice from the pot on the pork.

Tips:

1. Blanching the pork cleans the pork as well as making it easier to take on the color of the caramel.

2. Take care not to overcook the caramel as it burns easily.

3. Red yeast rice gives the pork a good red color making it more appetizing.

4. It is better to clean the bristles on skin of the pork belly by scraping with a knife rather than by burning because fire will cause the skin to shrink.

份量 6-8 人份
準備時間 30 分鐘
烹調時間 1 小時 15 分鐘

筍乾燜豬肉

◇

Braised Pork with Dried Bamboo Shoots

　　杭州臨安縣的天目山，與安徽黃山是同一山脈。位於天目山腳的黃坂鄉，是著名的竹筍產區，這裏出產的黃泥筍，筍肉壁厚，細嫩多汁，是杭州菜中重要的風味食材之一。

　　竹筍盛產於中國的東南、西南各省，以及嶺南地區；但南方沿海地區出產的竹筍，通常會帶有澀味，所以廣東人說筍是發物、毒物，不宜多食，其實這是指在與廣東潮濕溫暖的亞熱帶氣候所出產的竹筍。在嶺南以北江浙一帶的山區，冬天比較冷，春筍經過霜打雪凍才破土而出，這時筍肉已完成糖化，味道鮮美，而且沒有澀味。

　　竹筍的質量高低，除了氣候溫度的影響外，亦很大的因素在於生長的土壤。竹筍一般生長在黃泥土、黑泥土和灰泥土三種土壤，其中以生長在黃泥土的竹筍質量最佳，以至有不良商人，在竹筍上抹黃泥以冒充黃泥筍。而杭州天目山的竹筍，就是真正在黃泥土中種植的竹筍。

　　每年吃春筍的最佳時期，就是在過農曆年前後，過了三月份出產的春筍，已開始發青變老，這些老筍最適合加工成筍乾，以便於保存及運輸。筍乾是用鹽把切成條狀的筍醃好，再放在太陽下曬乾而成，亦有用烘焙的方法使之脫水而成。杭州菜中以筍乾做的傳統菜式很多，例如筍乾老鴨湯，筍乾冬瓜煮鹹肉和以下介紹的筍乾燜豬肉。

筍乾

● 材料

連皮五花肉 400 克	花椒 1 茶匙
薑汁 2 湯匙	薑粒 20 克
葱 2 條（切段）	蒜頭 8 瓣（切片）
老抽 1 湯匙	小紅椒 1 隻（切 3 段）
筍乾 150 克	紹興酒 60 毫升
紅麴米 1 茶匙	糖 1 茶匙
八角 1 粒	鹽（調味用）

● 做法

1. 把五花肉洗淨後瀝乾水分，切成約 2 厘米塊，放在大碗中，加入薑汁、葱段和老抽，醃半小時。

2. 把筍乾用清水洗去表面鹽分，再放入約 40℃暖水中，用手輕輕搓揉約 5 至 10 分鐘，中途可換暖水一次，至筍乾稍為變軟，撈出，用水沖洗，再瀝乾水分。先把筍乾較硬的節位部份切除不要，再切成約 2 厘米段。

3. 用香料袋把紅麴米、八角和花椒裝好，備用。

4. 在砂鍋裏燒熱 3 湯匙油，爆香薑粒和蒜片，放入豬肉，用中火炒 2 至 3 分鐘。

5. 放入小紅椒、紹興酒、醃汁和香料袋，再加水至完全覆蓋豬肉，大火煮沸，轉小火，加蓋燜約 45 分鐘。

6. 放入筍乾，加蓋再燜 45 分鐘，取出香料袋，加糖，再用大火收汁。試味後可加鹽調味。

● 烹調心得

1. 筍節比較硬，難以消化，建議把所有的筍節切除。

2. 筍乾含鹽分較高，在煮的時候先不要放鹽，到收汁時要試味，有需要時才加鹽調味。

3. 加入 1 隻小紅椒，是為了增加味道的層次，不愛吃辣就可以取消。

Serves: 6-8 / Preparation time: 30 minutes / Cooking time: 1 hour 15 minutes

Ingredients

400 g pork belly with skin

2 tbsp ginger juice

2 scallion, sectioned

1 tbsp dark soy sauce

150 g dried bamboo shoots

1 tsp red yeast rice

1 star anise

1 tsp Sichuan pepper

20 g chopped ginger

8 cloves garlic,, sliced

1 red chili pepper, cut in 3

60 ml Shaoxing wine

1 tsp sugar

salt for flavoring

Method

1. Rinse pork and cut into 2 cm pieces. Put into a large bowl and marinate with ginger juice, scallion and soy sauce for half hour.

2. Rinse dried bamboo shoots and soak in warm water (about 40°C). Rub gently for 5 to 10 minutes until soft. Change warm water once. Rinse with fresh water and cut away the knots that remain very firm. Cut into 2 cm pieces.

3. Put red yeast rice, star anise and Sichuan pepper in a spice bag.

4. Heat 3 tablespoons of oil in a casserole, stir fry ginger and garlic, put in pork and stir fry 2 to 3 minutes over medium heat.

5. Put in red chili pepper, wine, marinade and spice bag, and add water to cover pork completely. Bring to a boil, reduce to low heat, cover and simmer for about 45 minutes.

6. Add bamboo shoots, cover, and simmer for another 45 minutes. Remove spice bag, add sugar, and reduce over high heat. Flavor with salt if necessary.

Tips:

1. The knots of dried bamboo shoots can be quite firm and should be removed if necessary.

2. Dried bamboo shoots are rather salty. Do not add salt to the dish without tasting first.

3. Red chili pepper can be omitted.

糖 醋 小 排 骨

Sweet and Sour Spareribs

　　甜甜酸酸的味道，是大多數中國人的至愛，幾乎全國的餐館都在做糖醋排骨，近年更流行做加入草莓、山楂、杏等水果的糖醋排骨。港澳地區和外國唐人街流行的生炒排骨及咕嚕肉，更是深受中外食客喜愛的經典菜式。

　　在杭州任何一家餐館，都有「糖醋小排骨」這一道菜，很受歡迎。「糖醋小排骨」跟粵菜中的「生炒排骨」的明顯分別，是「糖醋小排骨」並不放紅紅綠綠的圓椒，更不放黃黃的菠蘿塊。更重要的是，粵菜的「生炒排骨」其實並非「生炒」，而是把排骨先炸，再加配料和甜酸汁來炒成，典型的「快」、「靚」、「正」。

　　「糖醋小排骨」有不同的做法，一種是用粵菜餐館做法先炸後炒；上海人的做法是把生的帶骨肋排，加冰糖、醋和紅醬油，翻炒至熟，再用猛火收汁至稠，排骨外形成一層甜酸蜜汁即可；還有一種更簡單，用壓力鍋放入排骨調好味道，「由頭到尾」煮至夠腍，取出大火收汁即可。

　　「糖醋小排骨」流行於上海、杭州等地區，做得好的「糖醋小排骨」，用的醬油是上海的「紅醬油」，近乎廣東醬油的老抽和日本、台灣的濃口醬油，主要用來上色，還有一種是「白醬油」，近似廣東醬油的生抽、頭抽，以及日本台灣的淡口醬油，用來調味。各地市場上叫做「醬油」的醬料，所含成份很難說得清楚，味道很複雜，有偏鹹的，有偏甜的，更有是帶酸味的，總之是一瓶兼用上色及調味。如果家中沒有「醬油」，也可以用一半生抽兌一半老抽來臨時對付；但不要選用草菇老抽，因為份量控制得不好的話，煮出來的排骨顏色會深黑。以下介紹的，是我們家的改良做法，更適合在家烹調。

● 材料

帶骨肋排 250 克	紅糖 4 湯匙
鹽 1/2 茶匙	紅醬油／老抽 1 湯匙
薑 20 克	鎮江香醋 2 湯匙
葱 4 條	生粉 1 茶匙

● 做法

1. 帶骨肋排切成約 2.5 厘米方塊，洗淨，用鹽和 4 湯匙清水拌勻，醃 15 分鐘。
2. 薑切成厚片稍拍裂，葱洗淨切段。
3. 將紅糖、紅醬油（或老抽）、和 1 湯匙鎮江香醋拌勻成調料，備用。
4. 把生粉拌入排骨中，燒熱 4 湯匙油，改小火，把排骨放入煎至微焦，取出，用清水洗去表面的油分。
5. 倒出鑊中的油，只留 1 湯匙，放下薑片和葱炒香，倒入排骨和調料，加 4 湯匙水，煮沸，用小火煮至收汁。
6. 用筷子挾走葱條，加入 1 湯匙鎮江香醋炒勻，即成。

● 烹調心得

1. 帶骨肋排切成約 2.5 厘米小方塊，塊頭可小不可大，是炒這道菜的重點，大塊的排骨難以快速上色、入味及炒熟。
2. 排骨先用鹽清水浸，作用是令肉質鬆嫩。

Serves: appetizer / **Preparation time:** 5 minutes

Cooking time: 15 minutes

Ingredients

250 g spareribs

1/2 tsp salt

20 g ginger

4 scallion

4 tbsp red sugar

1 tbsp red soy sauce / dark soy sauce

2 tbsp Zhenjiang vinegar

1 tsp corn starch

Method

1. Cut spareribs into 2.5 cm cubes, rinse, mix with salt and 4 tablespoons of water, and marinate for 15 minutes.
2. Cut ginger into thick slices and smash lightly. Cut scallions into sections.
3. Mix red sugar, soy sauce and 1 tablespoon of vinegar into a sauce.
4. Mix corn starch with spareribs. Heat 4 tablespoons of oil, and pan fry spareribs over low heat until slightly brown. Remove spareribs and rinse to rid of excess oil.
5. Pour out oil leaving about 1 tablespoon, stir fry ginger and scallions, add spareribs, sauce and 4 tablespoon of water, bring to a boil, and reduce to low heat to sauté until sauce is reduced.
6. Discard ginger and scallions, and stir in 1 tablespoon of vinegar.

Tips:

1. Spareribs should be cut small to easily take on color and flavor.
2. Pre-soak spareribs in salty water will tenderize the meat.

酒香炙骨頭

◆

Wine Flavored Spareribs

炙，（普通話：Zhi，粵語：隻），意思是把肉放在火上烤熟，烹調技術上屬火烹法，亦即今天的明火燒烤。現代已很少人用「炙」字，北方人流行叫「烤」，南方人通常叫做「燒」，所以叫做「南燒北烤」。

北宋徽宗皇帝趙佶（1028-1135），治國無能，最終客死異鄉，但他在書畫藝術上卻是才華橫溢，為後人留下了大量珍貴的書畫名作，特別是他的瘦金體書法，筆道瘦細而有力度，秀美灑脫而鋒芒畢露，獨創中國書法的一種超然境界。據說這道「炙骨頭」，是宋徽宗壽誕「天寧節」御宴上的第二盞御酒的下酒菜餚！宋室南遷後，這道菜成為臨安（今杭州）城中食肆酒家的名菜。

南宋周密《武林舊事》記載了「炙骨頭」這道宋室御用名菜，為了能更方便操作，我們把原來的明火燒烤，改為用焗爐（烤箱）來烤熟，即廣東話的「焗」。當然，如果你家中有明火燒烤的設備，那麼就可以做個原裝「宋版炙骨頭」，順便懷念一下這位苦命的大書法家宋徽宗！

份量　4 人份
準備時間　1 小時
烹調時間　45 分鐘

● 材料

豬肋排 600 克（切成約 7-8 厘米長）　　薑汁 2 湯匙

鹽 1.5 湯匙　　紹酒 3 湯匙

五香粉 1 茶匙　　海鮮醬 2 湯匙

花椒粉 1/2 湯匙　　麻油 2 湯匙

乾葱茸 2 湯匙　　桂花陳酒 90 毫升

　　桂花糖 1 湯匙

● 做法

1. 排骨洗淨後，用 1 湯匙鹽加 500 毫升清水浸泡 30 分鐘，
瀝乾。

2. 用 1/2 湯匙鹽、五香粉、花椒粉、乾葱茸、薑汁和紹酒
把排骨醃 30 分鐘。

3. 把海鮮醬用 2 湯匙水拌勻，再用麻油拌勻成甜醬，備用。

4. 把焗爐預熱到 180℃，在烤盤上放上排骨，烤 20 分鐘，
取出，再用桂花陳酒浸泡 15 分鐘。取出排骨，加甜醬拌
勻，放回烤盤內。

5. 把排骨放入焗爐上格，用燒烤火烤 5 分鐘，再把排骨翻
轉烤 5 分鐘。

6. 從焗爐中取出排骨，撒上桂花糖，盛碟，即成。

Serves: 4 / **Preparation time:** 1 hour / **Cooking time:** 45 minutes

Ingredients

600 g pork spareribs, cut into 7-8 cm length

1.5 tbsp salt

1 tsp five spice powder

1/2 tbsp ground Sichuan pepper

2 tbsp chopped shallot

2 tbsp ginger juice

3 tbsp Shaoxing wine

2 tbsp hoisin sauce

2 tbsp sesame oil

90 ml osmanthus flavored wine

1 tbsp osmanthus sugar

Method

1. Rinse spareribs and soak in 500 ml of water with 1 tablespoon of salt for 30 minutes. Drain.

2. Mix 1/2 tablespoon of salt, five spice powder, ground Sichuan pepper, chopped shallots, ginger juice and Shaoxing wine, and marinate spareribs for 30 minutes.

3. Mix hoisin sauce and 2 tablespoons of water, and stir in sesame oil to make a sweet sauce.

4. Pre-heat oven to 180°C, and roast spareribs for 20 minutes. Remove spareribs from the oven to soak in osmanthus flavored wine for 15 minutes. Take spareribs out of the wine mix with sweet sauce, and put on the baking tray.

5. Return spareribs to the oven and grill for 5 minutes. Turn over spareribs and grill for another 5 minutes.

6. Remove spareribs from the oven, brush with osmanthus sugar, and transfer to plate.

份量 6 人份
準備時間 20 分鐘
烹調時間 2.5 小時

醃篤鮮

Yanduxian Soup

　　杭州的冬天又濕又冷，一家人圍着吃飯，最窩心的就是來一個熱呼呼的砂鍋醃篤鮮。這個菜式據說起源於清代，醃篤鮮在杭州又叫做「南肉春筍」，因為杭州人把鹹肉叫做「南肉」。醃篤鮮是我國江南傳統的開春名菜，流行吃醃篤鮮的地區，包括上海、江蘇、浙江，難以準確地說是最早源自哪個地方。

　　醃篤鮮有三個主要材料：鹹肉、春筍、豬肉。清代著名文學家美食家李漁曰：「肉之肥者能甘，甘味入筍，則不見其甘，但覺味至鮮。」意思是認為煮筍必須配豬肉，而且要帶肥的豬肉，味道才會更鮮美。

　　竹筍，就是竹的嫩芽，每年農曆二、三月，正是春筍當造的季節，在杭州天目山竹林蔽天，出產的竹筍，嫩滑無渣爽脆鮮甜，是春筍中之極品。杭州的家鄉南肉（鹹肉）跟金華火腿一樣，都是用金華兩頭烏的豬肉醃製而成，但肉的部份不同，金華火腿用的是豬後腿，鹹肉是豬五花肉或肋條肉，金華火腿要醃製三年，而鹹肉醃製時間就短得多。

　　醃篤鮮這個名字其實一點也不古怪，而且對菜式的表達十分直接清楚，「醃」是代表鹹肉，「篤」是指這個湯菜是用長時間慢火燉，湯面在沸煮時發出「篤、篤、篤」的聲音，就像廣東方言說的：「滾到卜卜聲」，但也有人說「篤」字是指冬筍，因為兩個字都同為竹字頭。醃篤鮮的鮮就是鮮豬肉，再加上春筍、小棠菜這些時令新鮮蔬菜，就成了口味鹹鮮、湯白味濃、筍香鮮嫩、鹹鮮兩肉酥爛甘香的醃篤鮮。

材料

鹹肉 200 克

五花肉 200 克

薑 30 克

蔥 2 根

小棠菜 150 克

春筍 2 根

百頁結 200 克

小蘇打 1/2 茶匙

做法

1. 把鹹肉、五花肉汆水 3 分鐘，取出洗淨。

2. 薑切片，蔥打結。小棠菜修剪留嫩心部份，用滾水略燙，立即拿出用冷水沖洗，以保存菜的鮮綠顏色。

3. 沿春筍的長度，用刀把筍皮剖開，剝去筍皮，把根部老硬的部份切掉。筍尖空心的部份也同時去掉。用滾刀把筍切成塊，大火滾水汆燙 5 分鐘，撈出瀝水。

4. 百頁結用 1/2 茶匙小蘇打加約 500 毫升水泡 15 分鐘，拿出沖洗乾淨，同時擠出百頁結中心的水分，務求把所有小蘇打完全清除。

5. 把鹹肉、五花肉、薑和蔥放在砂鍋內，加水到完全覆蓋肉面為止。大火加蓋煮滾，轉小火煮 90 分鐘。

6. 加入春筍，再煮 45 分鐘。

7. 薑和蔥挾走不要，把鹹肉、五花腩取出切塊，和百頁結一同放回湯內，再煮 10 分鐘。

8. 最後放入小棠菜煮沸，試湯味調味，即成。

●烹調心得

1. 筍要預先汆水（出水），以去澀味。

2. 由於湯中有鹹肉，所以最後的調味要先試味才加鹽。

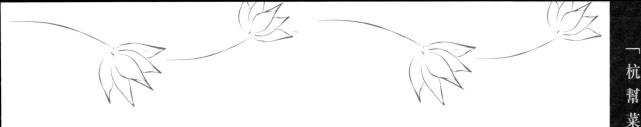

Serves: 6 / Preparation time: 20 minutes / Cooking time: 2.5 hours

Ingredients

200 g salted pork

200 g pork belly

30 g ginger

2 scallion

150 g Shanghai brassica

2 spring bamboo shoots

200 g knotted baiye

1/2 tsp baking soda

Method

1. Blanch salted pork and pork belly for 3 minutes, rinse.

2. Slice ginger and knot scallions. Trim and keep only the tender heart of the Shanghai brassica, blanch and rinse with cold water immediately to maintain their bright green color.

3. Slit open the bamboo shoots and remove the skins and cut off the hard part at the bottom of the shoots. Remove the hollow part of the shoots, and roll cut the remainder into chunk. Blanch for 5 minutes and drain.

4. Soak knotted baiye in 500 ml of water with 1/2 teaspoon of baking soda for 15 minutes. Rinse thoroughly and squeeze excess water to rid of all baking soda.

5. Place salted pork, pork belly, ginger and scallion in a casserole, and add water to cover all ingredients. Bring to a boil, reduce to low heat and cook for 90 minutes.

6. Add bamboo shoots and cook for another 45 minutes.

7. Discard ginger and scallion, cut salted pork and pork belly into pieces and return to the soup together with baiye. Boil another 10 minutes.

8. Add Shanghai brassica, sample and flavor with salt if necessary.

Tips:

1. Blanch bamboo shoots to remove any bitterness.

2. Taste soup before adding salt as much of the saltiness comes from the salted pork.

蜜汁鳳尾魚

Deep Fried Grenadier Anchovies

份量 8 人份
準備時間 30 分鐘
烹調時間 10 分鐘

上世紀六十年代，吃慣粵菜的香港人，開始流行吃外省菜，江浙菜館分高級的和平民化的兩種，平民化的江浙餐館一律被稱作「上海舖」，以配合所有江浙人都被香港本地人叫做「上海佬」，跟本分不清誰是上海人，誰是杭州人，誰是寧波人，那濃濃的鄉音，反正都不大聽得懂。店裏有賣北京填鴨、羊肉火鍋和鐵板鱸魚的是北京菜，其他的都當作是上海菜，其實當年香港也只有這兩種外省菜。

當時香港的「上海舖」多由華東各地來的新移民經營，一家大小胼手胝足的小生意，店面不大，價錢很大眾化，一大碗熱騰騰的上海麵，或者是被稱為客飯（一人套餐）的一碟小菜加兩碗白飯，還附送一碗湯，足以令人吃得很飽。那時候的大小「上海舖」食店，店前都有一個非常精彩的涼菜櫃枱，一大盤一大盤，展示着各式江浙冷盤前菜，有五香燻魚、油爆蝦、蜜汁鳳尾魚、雪裡蕻毛豆百葉、油燜筍、糖醋小排骨、油豆腐粉絲湯等等，客人可以隨意挑選。看得眼花撩亂，吃得心滿意足。時移世易，現在這些「上海舖」在香港已幾乎絕迹，看看滿街都是快餐店、意大利比薩、日本壽司和韓式炸雞，倒讓人十分懷念當年那些精彩的「上海舖」，還有那炸得香脆的蜜汁鳳尾魚。

鳳尾魚也叫做七絲鱭、鳳鱭、刀鱭，屬鯷科，肉味鮮，骨脆，最好的吃法是酥炸。幾十年前市場上已經有鳳尾魚的罐頭食品，暢銷中外。罐頭的鳳尾魚雖然也好吃，但是遠不如新鮮做的。這道菜其實不難做，只要在鳳尾魚上市的季節，花一點點時間，完全可以在家中做出香甜脆口的蜜汁鳳尾魚。

● 材料

鳳尾魚 600 克	蒜頭 2 瓣（剁碎）
鹽 1/2 茶匙	糖 3 湯匙
胡椒粉 1/8 茶匙	魚露 1 茶匙
薑茸 10 克	紹酒 1 湯匙

鳳尾魚

●做法

1. 鳳尾魚刮鱗①，切去魚頭②，同時用刀輕輕壓着魚頭往外拉，把魚內臟拉出③，洗淨。

2. 用鹽和胡椒粉拌勻鳳尾魚④，醃20分鐘，用廚紙吸去水分，抹乾⑤。

3. 大火燒熱炸油750毫升，把鳳尾魚炸至脆身⑥，取出備用。

4. 倒出鑊內油，只留1湯匙油燒熱，爆香薑蓉、蒜蓉，加糖、魚露和2湯匙清水、煮至汁稠⑦。

5. 放下鳳尾魚同炒，灒上紹酒，炒至收汁⑧。

●烹調心得

1. 鳳尾魚體形細小，用大火在最短時間內炸脆。火候如果不夠猛，鳳尾魚就炸不脆。

2. 香港每年春天，魚檔會有鳳尾魚出售，若有緣見到就不要錯過。

3. 市面上可以買到冷藏的鳳尾魚，要泡在冰水中解凍。

①

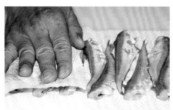

③

⑤

⑦

②

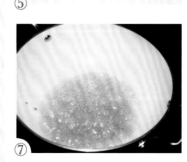

④

⑥

⑧

Serves: 8 / Preparation time: 30 minutes / Cooking time: 10 minutes

Ingredients

600 g grenadier anchovy

1/2 tsp salt

1/8 tsp ground white pepper

10 g ginger, grated

2 cloves garlic, chopped

3 tbsp sugar

1 tsp fish sauce

1 tbsp Shaoxing wine

Method

1. Scale fish ① , cut off fish head ② and gently pull to remove stomach ③ . Rinse and drain.

2. Marinate fish with salt and ground white pepper ④ for 20 minutes, pat dry with kitchen towel ⑤ .

3. Heat 750 ml of oil in a wok over high heat to a medium high temperature and deep fry fish until crispy ⑥ . Remove fish from the wok.

4. Pour out oil leaving only 1 tablespoon in the wok, stir fry ginger and garlic until pungent. Add sugar, fish sauce and 2 tablespoons of water and cook until sauce thickens ⑦ .

5. Put in fish and sprinkle wine along the inside of the wok. Stir gently until all the sauce adheres to fish ⑧ .

Tips:

1. High heat is required to deep fry grenadier anchovies to crunchiness.

2. Grenadier anchovies are available in the fish markets in the spring.

3. Frozen grenadier anchovies are available year round.

燻鯧魚

Spiced Pomfret

　　浙江東臨東海，海產豐富，盛產鯧魚、黃花魚和帶魚，鯧魚因體態扁平，北方人稱為「平魚」。燻鯧魚是浙江菜中常見的菜式，也有稱為「酥鯧魚」或「煙鯧魚」，是一道溫食的前菜。

　　「五香燻魚」或「蘇式燻魚」，在江蘇省和上海、港澳地區都很流行，這種「燻」，是江南菜中獨特的烹調法，而且只是做燻魚，不用於其他食材。做法就是把魚切片或塊，用油炸至焦脆，趁熱浸入事先調製好的醬汁中，使醬汁的味道迅速滲入魚肉中，成為外面酥脆，內裏鮮嫩的口感，別具風味。通常會用草魚／鯇魚來做，而杭州菜中的燻鯧魚，是一道上宴席的前菜，味道比起用草魚做的燻魚，味道更甘香鮮美，也更顯矜貴。

　　中菜烹飪法中的「燻」，正確來說是將醃過的生料或熟料，用米、麵粉、糖、蔗渣、茶葉、樟葉、松葉、果木等煙燻料，通過煙燻爐產生濃煙，把生或熟的原料燻至熟或燻至着色及入味，例如燻雞、燻蛋等菜式。但是，在烹調中並沒有經過煙燻過程，魚肉也沒有煙燻味，為甚麼稱為燻魚呢？原來在很久以前，燻魚的做法是把炸酥了的魚塊放入醬汁中吸味之後，再置於鐵絲網上，用白炭來燻烤，並不斷刷上醬汁，使其顏色一致，醬汁也稍乾成焦糖，這才成為真正的燻魚，不過這種做法現在已基本上很少人會做了。

●材料

鯧魚 1 條（約 450 克）	五香粉 1/2 湯匙
葱 2 根	鎮江醋 1.5 湯匙
薑茸 1 湯匙	片糖 1.5 塊（75 克）
鹽 1/2 茶匙	麻油 1.5 湯匙
紹興酒 3 湯匙	水 75 毫升

份量　4 人份
準備時間　15 分鐘
醃製時間　1 小時
烹調時間　15 分鐘

●做法

1. 鯧魚宰好洗淨，橫切去頭尾留用。魚身橫切成約
   3 厘米段，再把每一段切成三件。

2. 蔥白的部份切段拍扁，放入大碗內，加入薑茸、
   鹽和 1 湯匙紹興酒，把鯧魚塊拌勻，要確保每一塊
   魚都沾上醃料，醃製 1 小時。

3. 在另外的小鍋加入餘下 2 湯匙紹興酒、老抽、五香粉、鎮江醋、水和壓
   碎的片糖，再加入醃魚的汁和蔥青的部份，用小火煮至片糖完全溶化，
   用大碗盛起。

4. 鯧魚塊醃好後，用廚紙輕輕吸乾醬汁，用 1 公升炸油中火分批炸至金黃
   色，撈出後瀝去多餘的油分。

5. 趁熱逐一把剛炸好的魚塊放入蘸汁中稍浸一下，吸收汁味，然後拿出放
   在碟上，排回魚的形狀，再用毛筆掃上麻油，即可上桌。

●烹調心得

1. 如果愛吃炸脆的魚，可把炸好的魚塊，再用大火燒至高溫的炸油中，
   翻炸一下，再撈起瀝油，這才放入汁中。

2. 以片糖代替白糖，使燻魚更別具風味，也可以用 75 克白糖代替。

Serves: 4 / Preparation time: 15 minutes / Marinating time: 1 hour

Cooking time: 15 minutes

Ingredients

1 pomfret (about 450 g)

2 stalks scallion

1 tbsp, grated ginger

1/2 tsp salt

3 tbsp Shaoxing wine

1.5 tbsp dark soy sauce

1/2 tbsp five spice powder

1.5 tbsp Zhenjiang vinegar

1.5 pc pressed sugar (75 g)

1.5 tbsp sesame oil

75 ml water

Method

1. Clean and pat dry pomfret. Separate and save the head and tail. Cut the body crosswise into 3 cm sections, and each section into 3 pieces.

2. Cut scallion stems into sections and smash, put in a large bowl together with grated ginger, salt and 1 tablespoon of wine. Coat all the fish pieces including head and tail completely with sauce and marinate for 1 hour.

3. In another small pot, put in the 2 tablespoons of wine, soy sauce, five spice powder, vinegar, pressed sugar and water, and add the marinating sauce from the fish and scallion greens. Heat the sauce in low heat and cook until the pressed sugar has melted, then pour into a large bowl.

4. Pat dry fish fillet with kitchen towels and deep fry in 1 liter of oil until light brown. Remove and drain excess oil from the fish.

5. Dip fish while still hot into the sauce in the large bowl one at a time and coat completely with the sauce, then place on a plate and brush with sesame oil.

Tips:

1. If more crispness is desired, reheat the oil and deep fry fish a second time before dipping into the sauce.

2. Pressed sugar can be replaced by 75 g of white sugar.

蛤蜊汆鯽魚

Crucian Carp with Clams

蛤蜊指的是蚌類，包括各種大蜆小蜆，例如花蛤（花甲）、文蛤、桂花蚌等，品種繁多，盛產於中國東南沿海。蛤蜊自古就是人們盤中佳餚，以鮮味稱著，歷代古籍如《山海經》、《禮記》、《齊民要術》等都有記載。古代江浙的文人雅士，對蛤蜊的鮮味喜愛之餘，更撰詩歌頌。北宋的尚書都官員外郎、著名詩人梅堯臣詩曰：「紫綠常為海錯珍，吳鄉傳入楚鄉新。尊前已奪蟹螯味，當日蓴羹枉對人。」南宋兵部員外郎、著名文人晁公溯，對蛤蜊的美味更是充滿思念，詩曰：「使我轉憶江湖月，水珍海錯那可忘。十年不見尚能說，楚人未數鯉與魴。蛤蜊含漿自有味，蟹螯斫雪仍無腸。」如此描述，確令人對蛤蜊的鮮美垂涎三尺。

公元前 210 年，秦始王東巡到會稽（今紹興市）祭祀大禹，船隊在波濤洶湧的錢塘江無法前行，只能西行百餘里到今杭州市富陽區，錢塘江水系的富春江，在此江面狹窄處渡江去會稽，《史記・秦始皇本紀》中記載：「過丹陽，至錢塘，臨浙江，水波惡，乃西二十里，從狹中渡。」杭州地區的湖泊江河與錢塘江相連，而錢塘江外出東海，江、海、湖相通，水產相通，「蛤蜊汆鯽魚」正是海河水產合一的杭幫名菜。這道菜吃的是鮮味的蛤蜊肉和喝湯，鯽魚肉是可吃可不吃，但聽杭州朋友說，煮過湯的鯽魚頭香滑鮮美，是不少人的至愛美食。

「汆」（音川），意思與「淖」（音作）相同，是中菜烹調法的一種，流行於中國沿海地區，例如浙江菜、福建菜和廣東菜。「汆」的意思是把材料放在燒沸的湯或水中，快速焯（燙）至僅熟，以保持鮮味，常用於烹調海產，或者是比較容易熟的動物內臟，例如豬肝和豬腰。「汆」的另一種作用就是常常使用的「飛水」（出水），肉類在炒、煮湯或燜煮之前汆水以去掉血污羶味，或者除去海產的腥味，也有使切好的食材起到定型的作用，例如剞了花紋的魷魚。

●材料

花蛤／大蚌 600 克	鯽魚 1 或 2 條（約 600 克）	
小棠菜 50 克	紹酒 2 湯匙	
蔥 3 條	薑 20 克（切片）	
芫荽 1 棵	煮沸的水 500 毫升	
白胡椒粒 1/2 茶匙	鹽 1 茶匙	

●蘸料

薑絲約 1 茶匙
鎮江醋 1 湯匙

做法

1. 花蛤用清水加少許鹽和油浸 2 小時，吐淨泥沙，沖洗後瀝乾水分。

2. 燒熱白鑊，不要放油，放下花蛤，烘熱片刻，把遇熱打開殼的花蛤拿出，掰開已打開口的花蛤，殼的上蓋不要。不打開殼的要丟掉不能用。

3. 小棠菜洗淨，焯熟後備用。

4. 葱洗淨打成葱結，切碎芫荽，白胡椒粒稍為壓裂，備用。

5. 把蘸料的薑絲和醋混合成薑絲醋蘸料。

6. 鯽魚去鱗，翻開肚腔把血管的積血擠出，刮去黑色肚衣，沖洗乾淨。把魚脊的兩面從頭到尾各切一刀，刀深至骨。

7. 在鑊中下 2 湯匙油，燒至中低溫，放入鯽魚，中火煎約 3 分鐘，把魚翻轉再煎約 3 分鐘，轉大火，潷酒，放入薑片、葱結和白胡椒粒，加入煮沸的水，大火沸煮約 12 分鐘至湯變白色，加鹽調味。

8. 撈起鯽魚，放在大碗中，挾出薑和葱不要。小棠菜排在魚的兩邊。

9. 大火再煮沸魚湯，把花蛤放入煮 15 秒，取出花蛤排在碗上，把湯盛滿，撒上芫荽碎，即成。吃時拌以薑絲醋。

●烹調心得

1. 把花蛤放在熱鍋上烤，活的花蛤打開殼，已死的花蛤不會開殼。

2. 煮魚湯加入的水一定是用大沸水（滾水），而且全程用大火，否則溫度下降，煮出來的湯就不會是白色。

Ingredients

600 g large clams

50 g Shanghai brassica

3 scallion

1 coriander

1/2 tsp white peppercorn

600 g crucian carp (1 or 2)

2 tbsp Shaoxing wine

20 g sliced ginger

500 ml boiling water

1 tsp salt

Dip sauce ingredients

1 tsp shredded ginger

1 tbsp Zhenjiang vinegar

Method

1. Add a pinch of salt and a dash of oil to a pot of water and soak clams for 2 hours to allow them to regurgitate any sand. Rinse and drain.

2. Heat a dry wok, put in clams and roast until the clams open up. Remove the clams that are opened and pry open the shells completely. Discard the top half of the shells. Clams that do not open should be discarded.

3. Rinse and blanch Shanghai brassica.

4. Knot scallions and chop coriander. Crush white peppercorns.

5. Mix shredded ginger and vinegar into a dip sauce.

6. Press on the blood vessel along the spine on the inside of the fish to clear the blood and tear off the black membrane on the stomach. Rinse and clean thoroughly. Make a deep cut to the bone from head to tail along the back on each side of the fish.

7. Heat 2 tablespoons of oil in a wok to medium low temperature and pan fry fish over medium heat for about 3 minutes, turn over fish and pan fry for another 3 minutes. Change to high heat, sprinkle wine, add ginger, scallion and white peppercorn, and put in boiling water. Boil over high heat for about 12 minutes until soup turn white. Flavor with salt.

8. Transfer fish to a large bowl and put in Shanghai brassica. Discard ginger and scallions.

9. Re-boil fish soup over high heat, put in clams and cook for about 15 seconds. Remove clams to the bowl, add soup and top with chopped coriander. Serve with the dip sauce.

Tips:

1. Live clams will open their shells when roasted. Discard those that do not open.

2. To obtain a milky color soup, boiling water must be used and has to be cooked over high heat throughout the entire time.

份量 4 人份
準備時間 20 分鐘
烹調時間 15 分鐘

酒蒸石首

Steamed Corvina Fish in Wine

石首魚即黃魚，廣東人稱為黃花魚；因魚頭裏有兩粒耳石，故名；而大黃魚、小黃魚、梅童魚、獅子魚都屬於石首魚科。黃魚主要產於東海及東南沿岸，浙江舟山漁場所產的野生東海黃魚最為著名。野生黃魚生長期長，味道非常鮮美，肉質緊致嫩滑，千百年來深受人們喜愛。可惜因為環境的污染和過度的捕撈，野生黃魚產量急劇下降，現在市面上絕大多數的黃魚都是養殖的，味道遠遠比不上野生黃魚。

「酒蒸石首」即酒蒸黃魚，是一道帶湯的魚饌。根據南宋吳自牧《夢粱錄》中記載，是以香醇的黃酒來蒸黃魚，酒味香醇，魚肉鮮嫩，吃時以薑茸或薑絲加醋作為蘸料，風味獨特，清淡而不失矜貴。這道南宋時期的名菜，比起在清代姚燮（二石生）的《十洲春語》中記載的「雪菜大湯黃魚」，還早了好幾百年。

清洗黃花魚的方法

黃花魚腹部肉薄，不宜切破肚子，適宜用「壺抽法」處理，這是一種清除魚內臟而不用開膛破腹的方法，在我們陳家廚坊文化食譜系列的《天天吃海魚》一書中有詳細介紹。

1. 先把魚打鱗，割除魚鰓。
2. 在魚的肛門橫切一刀，把腸臟割斷。
3. 從口腔插入兩根筷子一直到肛門，夾住腸臟後用力攪拌一下使脫離腹腔。
4. 從口腔中把腸臟拉出。
5. 從口腔灌水清理腹腔。

材料

大黃魚 1 條（約 400 克）　　　　糖 1 茶匙

薑汁 1 湯匙　　　　　　　　　　金華火腿片 25 克

鹽 1 茶匙　　　　　　　　　　　薑片 6 片

葱 1 根（切段）　　　　　　　　上湯或雞湯 150 毫升

筍肉 50 克　　　　　　　　　　幼葱絲約 10 克

紹興酒 150 毫升

蘸料

薑絲 1 茶匙混合 3 湯匙鎮江醋

做法

1. 大黃魚洗淨，魚身兩面各斜刴 4 至 5 刀，用廚紙吸乾水分，以薑汁和鹽抹勻魚身內外，葱段放入魚肚中，醃 15 分鐘。

2. 筍肉切薄片，用水焯過，備用。

3. 紹興酒加糖拌勻。

4. 瀝去水分，把魚放入深蒸碟中，淋上紹興酒，在魚身上放上筍片和火腿片，在魚旁放上薑片，在碟邊注入上湯。

5. 用鋁箔紙（錫紙）密封，隔水蒸 15 分鐘，取出。

6. 揭去鋁箔紙，挾去薑片和葱結不要，放上葱絲，以薑絲醋為蘸料一起上桌，即成。

●烹調心得

這一道菜適合用香雪酒，是紹興酒的一個種類，有超過百年歷史，顏色淡黃透亮，味醇帶甜。香雪酒不容易買到，所以用普通紹興酒加少許糖代替，但不適宜用太雕酒，否則魚汁的顏色會過深。

Serves: 4 / **Preparation time:** 20 minutes / **Cooking time:** 15 minutes

---

Ingredients

1 corvina fish (about 400 g)

1 tbsp ginger juice

1 tsp salt

1 scallion, sectioned

50 g bamboo shoot

150 ml Shaoxing wine

1 tsp sugar

25 g Jinhua ham slices

6 ginger slices

150 ml chicken broth

10 g shredded scallion

Dip sauce ingredients

1 tsp shredded ginger

3 tbsp Zhenjiang vinegar

How to clean the fish

1. Clean the fish of scales and gills.
2. Make a cut at the anus at the end of the fish stomach to sever the intestines.
3. Insert two chopsticks through the head into the stomach through to the anus, grab hold of the fish maw and intestines, and twist to sever the intestines completely from the body.
4. Pull out intestines.
5. Flush the inside of the stomach thoroughly.

Method

1. Clean fish, and make 4 to 5 cuts on each side of the fish. Pat dry with kitchen towels and rub ginger juice and salt inside and out. Stuff scallion inside the fish stomach. Set aside for 15 minutes.
2. Cut bamboo shoot into slices and blanch.
3. Mix wine and sugar.
4. Drain fish, and put into a deep plate. Drizzle wine on the fish and top with bamboo shoot and ham. Place ginger slices next to the fish and add chicken broth from the side of the plate.
5. Seal plate with aluminum foil and steam for 15 minutes.
6. Lift the aluminum foil, discard ginger and scallions, and top with shredded scallions. Serve with the dip sauce.

Tips:

The proper wine to use with this dish is Xiangxue wine, a sweet yellow wine which can be replaced by add sugar to Shaoxing wine. Do not use Taidiao wine as the color is considered too dark for this dish.

份量　前菜小碟

準備時間　10 分鐘

烹調時間　20 分鐘

●材料

糟青魚乾 1 塊

　（約 300 克）

紹興酒 5 湯匙

白糖 1 湯匙

薑絲 20 克

●蘸料

薑蓉 1 湯匙

鎮江醋 4 湯匙

白糖 1 茶匙

●做法

1. 糟青魚乾用清水沖淨，瀝乾，蒸 10 分鐘，取出。

2. 把魚皮連鱗整塊撕去，再把魚肉撕成粗絲，排在蒸碟上，。

3. 把紹興酒和糖混合，淋在糟青魚絲上，放上薑絲，隔水蒸 10 分鐘，即成。

4. 吃時蘸薑蓉、鎮江醋和白糖混合的蘸料。

酒蒸糟青魚乾

Steamed Dried Black Carp

青魚是中國淡水養殖的四大家魚，廣東人稱為黑鯇，而香港人常吃的鯇魚是白鯇，又稱草魚。這兩種鯇魚一年四季都有供應，蛋白質豐富，骨刺少，可以清蒸、水煮、蔥㸆、紅燒，老少咸宜。

江南地區養殖的青魚，喜歡吃湖中的螺螄，肉質鮮美細嫩，杭州人俗稱當地的青魚為「螺螄青」。糟青魚乾是杭州著名的土特產，每年冬至前後，陽光充足，但氣溫下降，青魚肥美當造，這正是曬青魚乾的好時候，到初春時加入酒糟醃製，這樣到夏季就可以吃了。

糟青魚乾很容易保存，而且吃法很簡單，一般都會加薑絲和紹興酒來蒸，魚肉緊實，糟味香濃，沒有腥味，用來佐飯下酒都適宜，是傳統的杭州家常菜式。在香港要買杭州糟青魚乾，可以去南貨店購買。

糟青魚乾

Serves: appetizer / Preparation time: 10 minutes
Cooking time: 20 minutes

● Ingredients

300 g dried black carp

5 tbsp Shaoxing wine

1 tbsp sugar

20 g shredded ginger

● Dip sauce ingredients

1 tbsp, grated ginger

4 tbsp Zhenjiang vinegar

1 tsp sugar

● Method

1. Rinse fish, drain, and steam for 10 minutes.

2. Peel off and discard fish skin, tear flesh into thick strands and put on a plate.

3. Mix wine with sugar, drizzle on the fish and top with shredded ginger. Steam for 10 minutes.

4. Serve with dip sauce.

西 湖 醋 魚

Fish with Vinegar in West Lake Style

份量　6-8 人份
準備時間　5 分鐘
烹調時間　20 分鐘

　　西湖醋魚是杭州名菜,是江浙菜館必有的菜式。這個菜以前還有一個名字叫「叔嫂傳珍」。相傳古時有宋姓兄弟兩人,都很有學問,隱居在西湖以打魚為生。當地惡棍趙大官人有一次遊湖,見到一個在湖畔浣紗的婦人,見其面貌娟好,就想霸佔。知道這個婦人是宋兄之妻,就施用陰謀手段,害死了宋兄。宋家叔嫂非常悲憤,兩人一起上官府告狀,求官府主持公道,使惡棍受到懲罰。誰知官府勾結惡勢力,不但沒受理他們的告狀,反遭一頓棒打,把他們趕走。

　　回家後,宋嫂叫宋弟趕快收拾行裝外逃,以免惡棍前來報復。臨行前,嫂嫂燒了一盤魚,碗裏一滴油也沒有,而且加糖加醋,煮法奇特。宋弟好奇問嫂嫂,嫂嫂說:魚味有甜有酸,我是想讓你這次遠走他方,千萬不要忘記你哥哥是怎樣死的,你若將來生活得很好,也不要忘記老百姓受欺凌和你嫂嫂飲恨的辛酸。弟弟聽了很是激動,吃了魚,牢記嫂嫂的心意而去;後來,宋弟取得了功名回到杭州,報了殺兄之仇,把那個惡棍懲辦了。可這時宋嫂已經被逼離開了家鄉,宋弟一直尋找不到。有一次,宋弟出去赴宴,席間吃到一味魚,味道就是他離家時嫂嫂煮的一樣,連忙追問這是誰人煮的,才知道煮魚的正是他的嫂嫂。原來,嫂嫂為了避免惡棍來糾纏,隱名埋姓,替人做廚工。宋弟找到了嫂嫂,很是高興,便辭了官職,把嫂嫂接回家,重新以捕魚為生。

　　西湖醋魚整道菜不放油,但魚肉嫩滑,酸甜適中。我們採用浸煮的方法,就可以達到魚肉嫩滑的效果。香港市場上的鯇魚,一般都是大魚,買整條魚的話,份量嫌太多,而且家庭用的鑊尺寸比較小,難以浸熟,建議可買半邊的魚腩部份來代替整條魚。

材料

鯇魚 600 克

薑片 20 克

鹽 1 茶匙

鎮江醋 50 毫升

薑絲 20 克

紅糖 2 茶匙

生粉 1.5 湯匙

做法

1. 鯇魚宰好去鱗，用小刀輕輕刮去腹腔內的黑膜，清洗乾淨。

2. 在鑊中注滿八成水，放進薑片和 1/2 茶匙鹽，大火煮至沸，熄火。

3. 把鯇魚慢慢放進鑊裏，皮朝上浮在水中，蓋上鑊蓋 12 至 14 分鐘至把魚浸熟。

4. 把魚小心輕輕撈起放在碟中，倒出碟中的水。

5. 倒出鑊中的水，只留約 200 毫升，放入醋、薑絲、糖和 1/2 茶匙鹽，煮沸。

6. 用生粉開 2 湯匙水攪勻，除除倒進小鍋中勾芡。

7. 把糖醋汁淋在魚上，即可上桌。

●烹調心得

西湖醋魚的要求是不放油，但魚肉十分嫩滑。浸魚要熄火浸才會嫩滑。水溫度過高或不斷沸騰，就會使魚肉變得粗糙。如果魚身較厚，可在浸 10 分鐘後再開小火提升水溫。

Serves: 6-8 / **Preparation time:** 5 minutes / **Cooking time:** 20 minutes

 Ingredients

600 g grass carp

20 g ginger slices

1 tsp salt

50 ml Zhenjiang vinegar

20 g shredded ginger

2 tsp red sugar

1.5 tbsp corn starch

Method

1. Scale and clean the fish, scrape clean the black membrane from inside the fish using a small knife.

2. Fill up a wok to 80% with water, add ginger slices, 1/2 teaspoon of salt, bring to a boil and turn off the heat.

3. Lower the fish skin side up into the water, cover and coddle for 12 to 14 minutes until the fish is cooked.

4. Carefully lift the fish onto a plate, and then drain water from the plate.

5. Pour out most of the water from the wok leaving about 200 ml, add vinegar, shredded ginger, sugar and 1/2 teaspoon of salt, and bring to a boil.

6. Mix corn starch with 2 tablespoons of water and stir into the water to thicken into a sweet and sour sauce.

7. Pour sauce over the fish before serving.

Tips:

The flesh of the fish should be tender and soft even without adding any oil. The fish should be steeped only in hot water that has been just boiled but without additional heat. For a large fish, heat may be added after about 10 minutes to maintain the proper temperature of the water.

份量 4-6人份

準備時間 20分鐘

烹調時間 25分鐘

砂 鍋 魚 頭 豆 腐

Fish Head with Tofu in a Casserole

　　乾隆六下江南，留下了不少風流傳說，也為後人提供了很多與飲食相關的趣事。話說乾隆微服出巡到訪吳山（在今天杭州市內），走到半山腰卻下起大雨，乾隆饑寒交迫，便走進一戶山野人家希望找一些食品充饑。屋主王潤興是個食物小販，見來人如此狼狽模樣，頓生同情心，便把當日沒賣出去的一個魚頭，加上一些青菜和豆腐，放入一些醬料，用一個破砂鍋燉好給乾隆吃，乾隆吃得狼吞虎嚥，覺得這菜比宮殿中的任何山珍海味都更美味。

　　後來乾隆再訪吳山，他沒忘記當日吃的這頓美食，再次訪杭州時，到了當日的破屋，再找到王潤興，叫他開一個飯店賣魚頭豆腐，並當即賞賜他五百兩銀子，資助他在吳山腳下開了一家飯館，還提筆寫下「皇飯兒」三個大字，下款是乾隆御筆。王潤興這才知道他遇上了當今聖上。從此，王潤興便把乾隆御筆「皇飯兒」掛在中堂，專營魚頭燉豆腐和小菜，結果名利雙收。有人為此事題曰：「肚饑飯碗小，魚美酒腸寬，問客何所好，豆腐燒魚頭。」這是一個好心有好報的動人故事，這道普通不過的砂鍋魚頭豆腐，從此也有另一名字：「乾隆魚頭豆腐」。

　　「砂鍋魚頭豆腐」是一道傳統的著名菜式，在杭州用的是千島湖的「胖頭魚」（鱅魚）的魚頭，魚頭肥潤嫩滑，豆腐細嫩入味，湯汁醇厚。鱅魚頭不但甘腴味美，而且營養豐富，以大魚頭入藥，可治風寒頭痛。明代李時珍在《本草綱目》中曰：「鱅之美在於頭」，《本草求源》中說魚頭可以「暖胃，去頭眩，益腦髓，老人痰喘宜之」。鱅魚即廣東的大魚，是我國南方四大家魚之一，一年四季都很容易買到。

材料

大魚頭 1 個	板豆腐 2 塊	薑片 4 片
薑汁 1 湯匙	蒜頭 5 瓣	紹興酒 1 湯匙
胡椒粉 1/4 茶匙	乾冬菇 4 朵	糖 1/2 茶匙
豆瓣醬 2 湯匙	生粉 2 湯匙	生抽 1 茶匙
紹菜 300 克	清雞湯 500 毫升	鹽 1/2 茶匙

做法

1. 魚頭砍成兩邊，洗淨瀝乾後，內外抹上薑汁和胡椒粉。壓碎豆瓣醬，抹在魚頭上，醃 15 分鐘。

2. 紹菜洗淨，切成約 7 厘米長 x 2 厘米寬，用水灼熟，備用。

3. 豆腐切成 1 厘米厚塊，用水稍為焯過，備用。

4. 冬菇泡軟，去蒂，切厚片，備用。

5. 在砂鍋內燒熱 4 湯匙油，放入蒜頭炸至微黃，取出備用。

6. 把魚頭拍上生粉，放入油中煎至金黃，取出備用。

7. 倒出砂鍋內的油，只留 1 湯匙，爆香薑片、蒜頭和冬菇，放入清雞湯和紹菜，再把魚頭放在上面，加入紹興酒、糖、生抽和鹽，大火煮 10 分鐘，放入豆腐，煮 5 分鐘，原鍋上桌，即成。

●烹調心得

1. 上海豆瓣醬辣味不重，如果用四川豆瓣醬則味道較辣。不吃辣的話，可用廣東麵豉醬代替。

2. 無論用的是那一種醬，都容易沾鍋底，所以要小心留意，必要時可加少許沸水。

3. 也可以用清水代替雞湯，但注意要再加多半茶匙鹽。

Serves: 4-6 / **Preparation time:** 20 minutes / **Cooking time:** 25 minutes

Ingredients

1 bighead carp fish head	2 firm tofu	4 slices ginger
1 tbsp ginger juice	5 cloves garlic	1 tbsp Shaoxing wine
1/4 tsp ground white pepper	4 dried black mushrooms	1/2 tsp sugar
2 tbsp chili bean paste	2 tbsp corn starch	1 tsp light soy sauce
300 g Napa cabbage	500 ml chicken broth	1/2 tsp salt

Method

1. Cut fish head down the middle into two pieces, rinse, and rub thoroughly with ginger juice and ground white pepper. Crush chili bean paste and mix with fish head, and marinate for 15 minutes.
2. Rinse cabbage and cut into pieces 7 cm x 2 cm. Blanch.
3. Cut tofu into 1 cm thick pieces and blanch.
4. Soak mushrooms until soft, remove stem, and cut into thick slices.
5. Heat 4 tablespoons of oil in a casserole and deep fry garlic until light brown. Take out garlic.
6. Coat fish head with corn starch and pan fry until golden. Take out fish head.
7. Pour out oil leaving 1 tablespoon, and stir fry ginger, garlic and mushroom until fragrant. Put in chicken broth and cabbage, top with fish head, and add wine, sugar, soy sauce and salt and boil for 10 minutes over high heat. Put in tofu and cook for another 5 minutes before serving in the casserole.

Tips:

1. Shanghai chili bean paste is only mildly hot compared to Sichuan chili bean paste, but Cantonese bean paste, which is not hot, can also be used.
2. Bean paste can easily stick to the bottom of the casserole. Add a little boiling water if necessary.
3. Add an additional 1/2 teaspoon of salt if water is used in place of chicken broth.

龍井蝦仁

◆

Shrimps with Longjing Tea

　　杭州菜中的「龍井蝦仁」，以新鮮龍井茶代替青葱來炒河蝦仁，色澤亮麗清雅，是一道具代表性的杭州名菜。龍井茶是中國十大名茶之一，產於杭州地區，一千多年前唐代茶聖陸羽，撰寫了世界上第一本以茶作專題的書《茶經》，書中已經記載了杭州山區生產茶葉的資料。相傳唐代法欽禪師雲遊至徑山，見群山翠綠，流水淙淙，樵夫指路說：「此上徑山，徑通天目」，於是法欽禪師就決定在此地開山建寺。到了南宋，宋孝宗遊徑山，寫下了一塊御碑「徑山興聖萬壽禪寺」。南宋之後，徑山成為日本的佛教聖地，到中國學佛的日本僧人，都要到此禮佛，學習種茶和茶道。

　　青茶以「龍井茶」作為名字，是在宋朝開始，眾多著名文人以詩詞歌頌，更使品嘗龍井茶成為時尚。「白雲峰下兩旗新，膩綠長鮮谷雨春」，這兩句詩是北宋詩人蘇東坡對龍井茶的讚美，他更在杭州獅峰山的懸岩上，寫下了至今尚存的匾額「老龍井」。老龍井是一江青泉，舊名「龍泓」，與虎跑、玉泉並稱西湖三大名泉，泉水含豐富礦物質，用來泡茶自是一絕。每年清明前採摘的「明前龍井」（雨前龍井），更是極品名茶之首，曾是清乾隆皇帝指定的貢品茶葉。

　　取龍井新茶炒河蝦仁，粉紅色的蝦仁配上碧綠的茶葉，典雅高貴，味道鮮甜清香。此菜最好是選用鮮活河蝦取仁，也可以用冷藏河蝦仁代替，味道和口感雖然稍遜，但也值得一試。

份量　4 人份
準備時間　10 分鐘
冷凍時間　1 小時
烹調時間　5 分鐘

●材料

蝦仁 300 克

鹽 1/2 茶匙

雞蛋白 1/2 個

龍井茶 1 克

紹興酒 1 湯匙

●做法

1. 蝦仁洗淨瀝乾水分，用 1/4 茶匙鹽抓洗 1/2 分鐘，用清水沖淨鹽分。

2. 用廚紙把蝦仁的水分吸乾，再用乾淨毛巾捲起，放在冰箱裏冷凍 1 小時。

3. 取出蝦仁盛在碗內，加入 1/4 茶匙鹽和雞蛋白拌勻。

4. 在茶杯裏放入龍井茶葉，用沸水 50 毫升沏泡（不要加蓋）1 分鐘，把茶倒掉，剩下的茶葉留用。

5. 在鑊裏把 500 毫升油熱至低溫（約 120℃），放入蝦仁，迅速用筷子劃散（約 15 秒），立即取出，瀝油。把鑊裏的油倒出，留作他用。

6. 大火燒熱 2 茶匙油，倒入蝦仁爆炒至全熟，鍋邊潽紹酒炒勻，隨即把茶葉倒入，快炒至紹酒完全蒸發，即可上碟。

●烹調心得

1. 優質的河蝦仁不易買到，可以用冷藏的南美白蝦仁代替。

2. 解凍冷藏蝦仁，最好的方法是把蝦仁泡在冰水裏，讓蝦仁慢慢均勻地解凍，蝦仁肉就不會變霉。

Serves: 4 / Preparation time: 10 minutes / Refrigeration time: 1 hour / Cooking time: 5 minutes

Ingredients

300 g shelled shrimps

1/2 tsp salt

1/2 egg white

1 g Longjing tea

1 tbsp Shaoxing wine

Method

1. Rinse shrimps and drain. Rub with 1/4 teaspoon of salt for 1/2 minute, then rinse with fresh water.

2. Dry the shrimps with kitchen towels, roll up shrimps in a clean towel and refrigerate for 1 hour.

3. Place refrigerated shrimps in a bowl and mix with 1/4 teaspoon of salt and egg white.

4. Brew Longjing tea with 50 ml of boiling water for 1 minute and pour out the tea leaving tea leaves for later use.

5. Heat 500 ml of oil to a low temperature (about 120°C), add shrimps, stir with chopsticks for about 15 seconds and remove immediately to drain. Pour out oil and save for other uses.

6. Heat 2 teaspoon of oil in a wok over high heat, stir fry shrimps until fully cooked, drizzle wine along the inside of the wok, add tea leaves and stir fry rapidly until the wine evaporates. Transfer to plate.

Tips:

1. If shelled river shrimps are not available, shelled white shrimps can be used as a replacement.

2. The best way to thaw frozen is to soak it in iced water and allow them to thaw slowly.

油爆蝦

Shrimps Sautēed in oil

份量 4-6 人份
準備時間 15 分鐘
烹調時間 20 分鐘

做油爆蝦用的是淡水蝦，淡水蝦生長在河流、水庫或沼澤，常見的品種有淡水白蝦、泥蝦、羅氏沼蝦和淡水大頭蝦。淡水白蝦殼較薄，肉質甜美，適合做清炒蝦仁，淡水泥蝦的肉爽口，很適合做廣東點心中的蝦餃。羅氏沼蝦體形較大，蝦殼較硬，適合去殼做炒蝦球。而體形較小的淡水大頭蝦和淡水白蝦，蝦殼軟硬適中，最適宜做油爆蝦。海蝦因為要長期抵受密度高的海水帶來的壓力而長出厚的蝦殼，如果買不到淡水蝦，也可以用半淡水養殖的基圍蝦代替，但要選用體形較小的蝦。

做得好的油爆蝦，要脆得連蝦頭蝦殼都可以吃下，為了做到把蝦殼炸脆，而蝦肉不會過火收縮，方法是把蝦分三成幾批處理。由於生的蝦又涼又濕，如果把蝦全部一次過放入 220℃的炸油中，那麼油溫就會一下子降下去變成約 180℃，這樣便難以炸脆。所以一碟油爆蝦的份量，要分三至四批來炸。當一批蝦放入到 220℃的沸油中，油溫也會逐步下降，因此當看到炸油的泡減少，意味着油溫已經下降，就要立刻把蝦撈出離開炸油，等油溫再度升高至 220℃時，把蝦再放下去，經過這樣三次反復油炸後，保證可以達到香脆而蝦肉不過火。在酒家做這道菜，不需要這樣反復油炸，這是由於專業的爐火很旺，油溫度很高，而且還有大鍋炸油。在家裏只是家用爐具，更不想用太多的炸油，想吃香脆可口的油爆蝦，就要用多些心思和耐性來分批炸了。

● 材料

淡水蝦 400 克	生抽 1 茶匙
薑茸 1 湯匙	鎮江醋 1 湯匙
葱花 2 湯匙	紹酒 1 茶匙
糖 2 湯匙	麻油 1/2 茶匙
鹽 1/2 茶匙	油 1 公升

● 做法

1. 剪去蝦頭的尖刺部份和蝦腳，沖洗乾淨，用廚紙吸乾水。

2. 準備好一隻大漏勺，用大火加熱炸油至約 220℃，拿起三分一的蝦放入漏勺中，放進油中炸，見到油中泡沫少了，拿起漏勺，使蝦離開炸油，等油溫再升高後，把炸過的蝦再放進炸油中，以上方法重複兩次，即共三次，撈出蝦瀝油，再用廚紙吸收多餘的油份。餘下的蝦分兩批按以上程式炸脆。

3. 把炸蝦的油倒出，只留 2 湯匙油在鑊內。

4. 大火紅鑊，放入糖和蝦快手爆炒，再加入薑茸，立即在鑊邊濽紹酒，加入生抽、鹽和鎮江醋炒勻，再撒下葱花爆炒至收乾汁水，最後淋少許麻油炒勻，即可盛碟。

● 烹調心得

1. 在放醬油和醋之前，先放糖和炸好的蝦，立即快手爆炒，就是要在炸脆的蝦殼上一層焦糖，增加甜香味道之餘，可減低蝦殼受醬油和醋深顏色的影響，顏色會更紅亮一些。

2. 炒的時間越短蝦就越脆，而且不能打芡，否則蝦會立即變軟。

3. 蝦炸好後要隨即就進行爆炒的步驟，否則蝦頭放久了也會變暗色。

4. 濽酒要濽在鑊邊，不要直接淋到蝦上，以防火候不當，蝦肉可能變霉。

Ingredients

400 g river shrimps

1 tbsp, grated ginger

2 tbsp chopped scallion

2 tbsp sugar

1/2 tsp salt

1 tsp light soy sauce

1 tbsp Zhenjiang vinegar

1 tsp Shaoxing wine

1/2 tsp sesame oil

1 liter oil

Method

1. Using kitchen scissors, cut off the sharp claws from the shrimp's head and the legs from the body. Rinse and pat dry with kitchen towels.

2. Heat up the oil in the wok to about 220°C over high heat, put a handful of shrimps into a hand held metal colander, and lower into the oil to deep fry. When the bubbles of the oil begin to subside, lift the colander to remove the shrimps from the oil and wait until the oil is heated up again before putting the shrimp back in. Repeat the process twice (for a total of three times) for each handful of shrimps. Use the same procedure for the remainder of the shrimps.

3. Pour out the oil, leaving only 2 tablespoons of oil in the wok.

4. Heat up the wok under high heat, quickly stir fry the shrimps together with sugar, add grated ginger, sprinkle Shaoxing wine along the inside of the wok, then put in soy sauce, salt and vinegar, finally throw in chopped scallions and stir fry until all the sauces are dried up. Stir in sesame oil and transfer to plate.

Tips:

1. Stir frying shrimps and sugar before putting in soy sauce and vinegar is to allow a coating of caramel to be formed on the surface of the shrimps to enhance the color of the shrimps.

2. Keeping the stir frying time short will help to keep the crispness of the shrimps. Do not thicken with gravy.

3. Stir fry and serve the shrimps immediately after they have been deep fried. The color of the shrimp heads will darken with time.

4. Always sprinkle wine along the side of the wok. Avoid putting wine directly on the shrimps.

蟶子的廣東話發音是「勝子」，但普通話的「蟶」字為 cheng（音撐），所以廣東人到了外省吃蟶子，記得蟶子是「撐」子而不是「星」子。浙江沿海盛產蟶子，香港市場上也是長年有供應，只會因季節不同，蟶子的肉肥瘦有別。

　　記得前幾年的四月，與兩位好朋友遊杭州，卻遇上春雨矇矓的惱人天氣。微風細雨下，我們逛完絲綢市場，走進旁邊一家很普通的酒家吃午飯，隨便點了幾個杭州小菜，見水產缸中養着新鮮蟶子，便點了一碟三絲炒蟶子。誰知這碟炒蟶子，口感嫩滑，味道鮮甜，似在吃生蠔。立刻叫來服務員，請廚房再照辦炒一碟，結果又是一掃而光。我們算得上是嘗過無數美食，但值得急急再叫來一盤的菜餚，卻是僅此一次，印象難以忘懷。

　　蟶子即竹蚶，英文名字是 RazorClam，即剃刀蚶，有長、短、肥、瘦等不同的品種。蟶子生長在近海的泥地上，殼呈長方形，煮熟後肉質雪白嬌嫩，味道鮮美，肥的蟶子吃起來有點像生蠔。蟶子可蒸、焯、炒、炸和煮湯。不過，蟶子的美味，基本上蓄於蟶子肉的水分中，所以不能長時間煮，稍為過熟，鮮味就隨水分流走了，肉質也會有失嫩滑。

三絲炒蟶子

Stir Fried Razor Clams

份量　4 人份

準備時間　15 分鐘

浸泡時間　30 分鐘

烹調時間　5 分鐘

● 材料

新鮮蜆子 600 克	金華火腿 20 克
鹽 1 茶匙（浸泡用）	鹽 1/2 茶匙
薑汁 1 湯匙	紹興酒 1 湯匙
蒜頭 8 瓣	鎮江醋 1 茶匙
冬菇 5 朵	生抽 1 茶匙
青圓椒 1/2 個	麻油 1 茶匙

● 做法

1. 帶殼蜆子洗去泥污，放在水中浸 30 分鐘，水要浸過蜆子，並加入 1 茶匙鹽在水中攪溶。

2. 把蜆子撈出，煮沸一大鍋水，放入薑汁，把蜆子倒入，直至蜆子打開口，立即撈出，拆去殼取出蜆子肉，用廚紙吸乾水分。

3. 蒜頭去衣，中間切開成兩粒。

4. 冬菇浸透去蒂切絲，青圓椒切絲，金華火腿蒸 5 分鐘後取出切絲。

5. 燒熱 3 湯匙油，放入蒜粒爆香。放入冬菇絲和鹽，爆炒 1 分鐘。

6. 放入蜆子肉、青椒絲和火腿絲，灒紹酒，加入醋和生抽，快手爆炒約 1 分鐘，加麻油，即成。

● 烹調心得

1. 蜆子不要焯得過熟，剛開口就可以了。不開口的蜆子是不能吃的。

2. 拆蜆子肉的時候，把腸臟撕掉不要。

Serves: 4 / Preparation time: 15 minutes / Soaking time: 30 minutes

Cooking time: 5 minutes

Ingredients

600 g fresh razor clams

1 tsp salt (for soaking)

1 tbsp ginger juice

8 cloves garlic

5 dried black mushrooms

1/2 pc green bell pepper

20 g Jinhua ham

1/2 tsp salt

1 tbsp Shaoxing wine

1 tsp Zhenjiang vinegar

1 tsp light soy sauce

1 tsp sesame oil

Method

1. Rinse the clams and soak in fresh water for 30 minutes with 1 teaspoon of salt added.

2. Drain the clams. Bring a large pot of water to a boil, add ginger juice, and put in the clams. Once the clams open up, remove from the pot immediately and take out the meat. Soak up excess water on the meat with kitchen towels.

3. Peel garlic and cut each into two halves.

4. Soak mushrooms until soft, remove stem and cut into strips. Deseed bell pepper and cut into strips. Steam Jinhua ham for about 5 minutes and cut into strips.

5. Heat 3 tablespoons of oil, stir fry garlic until aroma begins to rise, add mushrooms and salt and stir fry for about 1 minute.

6. Put in clam meat, bell peppers and Jinhua ham, sprinkle wine, add vinegar and soy sauce and stir fry for about 1 minute. Finally stir in sesame oil before platting.

Tips:

1. When boiling the clams, cook just enough for them to open up. Unopened clams should not be eaten.

2. Be sure to remove the intestines of the clams when taking them out of the shells.

生爆鱔片

Stir Fried Eel Pieces

●材料

黃鱔 500 克	糖 1 茶匙	銀芽（綠豆芽）100 克
生粉 1 茶匙	肥豬肉粒 10 克	胡椒粉 1/2 茶匙
鹽（洗黃鱔用）1 茶匙	薑絲 10 克	蒜茸 2 湯匙
鹽 1/2 茶匙	紹興酒 1 茶匙	麻油 1 湯匙
生抽 2 茶匙	韭黃切段 100 克	

中國人食用黃鱔的歷史悠久，中醫認為黃鱔有補氣養血、袪風通絡、滋補肝腎的功效。黃鱔一年四季都有出產，但端午節後 1 個月的黃鱔最滋補，所以民間有「小暑黃鱔賽人參」之說。

這道「生爆鱔片」，以前叫做「生爆鱔背」，在我們的食譜書《在家做江浙菜》中曾介紹過。江浙地區傳統的宰鱔方法，是在活黃鱔的背上開刀，但現在只有上海一些菜市場還保留這種宰法。

「生爆鱔片」是採用北方蒜爆的技法，是典型的「南料北烹」菜式。「生爆」，顧名思義是把全生材料（或稍作瞬間氽水去血污），放入炒鍋中用猛火高油溫，快速爆炒至僅熟，吃起來就會爽脆鮮甜。不過，自上世紀七十年代開始，國內大部份酒家的做法，是把黃鱔片先上粉炸或掛粉糊後，油炸至脆，然後炒製埋芡汁而成，這是屬於炒法中的「焦炒」技法。焦炒的鱔片，由於經過加粉炸脆，鱔片炸後會鬆起，實際用鱔的份量就可以減少一些，而且對鱔肉的新鮮程度要求不高，不少酒家就會用冰鮮或冷藏的黃鱔肉，而不用活宰黃鱔，以降低成本，甚至在杭州的一些著名酒家，吃到的生爆鱔片都是焦炒的，這已是約定俗成。但與傳統生爆做的鱔片比較，口感和鮮味自然是不同。

● 烹調心得

1. 爆的意思是要火猛手快，不要加蓋，開始下鍋炒前，先用小碗把所有的調料準備好，炒的時候不會浪費時間，不要把鱔片炒得過熟，否則就會失去軟滑爽口的效果。用新鮮活宰的鱔片來做爆炒，不必勾芡，這菜式與炒鱔糊不同，生爆鱔片要求炒得清爽俐落，盤子底不應該有醬汁。

2. 韭黃見火易熟，不必擔心韭黃不熟，如果完全熟透口感就會韌。銀芽就是擇去豆頭豆尾的綠豆芽，在這道菜式中的作用是增加脆口感，也是一炒匀即起，銀芽一旦過熟就會出水，整盤菜就會失敗了；所以新手下廚如果怕失敗，也可以省去銀芽。

● 做法

1. 黃鱔活宰及去骨，頭和骨都不要，清除所有內臟①。黃鱔肉用生粉和鹽洗過，用水洗乾淨後瀝乾，橫切成每段 5-6 厘米②，再直切成約 2 厘米寬的鱔片③。

2. 大火煮水大沸騰後熄火，倒入鱔片④，用筷子迅速攪動，不要開火，目的在余去血水，10 秒鐘左右撈起，用清水沖洗後瀝水備用⑤。

3. 把鹽、醬油和糖的份量先量好，用一隻小碗裝好拌勻成醬料。

4. 用 1 茶匙油起鑊，倒入肥豬肉粒，炸出豬油後撈走豬油渣。

5. 大火燒熱豬油，放入薑絲和鱔片一起爆炒到鱔片乾身⑥，然後在鑊邊潷入紹興酒⑦，爆炒十多下後加入拌好的醬料⑧，快手爆炒到鱔片熟透，放入韭黃段和銀芽炒勻⑨，再加胡椒粉兜勻即上碟。

6. 用筷子把鱔片撥開，中間留一小凹位，在凹位中放入蒜茸⑩。

7. 用另外一隻乾淨鑊燒沸 1 湯匙生油和 1 湯匙麻油，淋在蒜茸堆上⑪，立刻上菜。要求在上菜時，蒜茸油仍熱得滋滋作響。

① ② ③
④ ⑤ ⑥
⑦ ⑧ ⑨
⑩ ⑪

Serves: 4-6 / Preparation time: 10 minutes **/ Cooking time:** 15 minutes

Ingredients

500 g yellow finless eel

1 tsp corn starch

1 tsp salt (for cleaning)

1/2 tsp salt

2 tsp light soy sauce

1 tsp sugar

10 g chopped fatty pork

10 g shredded ginger

1 tsp Shaoxing wine

100 g hotbed chives, sectioned

100 g bean sprouts

1/2 tsp ground white pepper

2 tbsp chopped garlic

1 tbsp sesame oil

Method

1. De-bone eel, discard head and bones, and remove intestines ① . Clean the eel with salt and corn starch, rinse and drain, and cut into sections 5-6 cm long ② and pieces of about 2 cm wide ③ .

2. Bring a large pot of water to a boil, and turn off the heat. Put in eel pieces ④ and stir rapidly for about 10 seconds, then rinse with cold water and drain ⑤ .

3. In a small bowl, mix salt, soy sauce and sugar into a sauce.

4. Heat up 1 teaspoon of oil in a wok, put in chopped fatty pork and fry until crispy. Remove the crispy pork fat.

5. Heat up the oil in the wok, put in shredded ginger and eel pieces and stir fry rapidly to rid of the moisture ⑥ . Sprinkle wine along the inside of wok ⑦ , stir fry the eel lightly and add the sauce ⑧ . Stir fry rapidly until eel pieces are thoroughly cooked, add hotbed chives and bean sprouts ⑨ and toss, put in ground white pepper, mix well and transfer to plate.

6. Move the eel pieces in the plate to create an opening in the center for the chopped garlic and add garlic ⑩

7. In a clean wok, heat up 1 tablespoon of oil and 1 tablespoon of sesame oil; pour over the chopped garlic ⑪ before serving.

Tips:

1. Stir fried eel has to be done very rapidly over high heat to avoid overcooking. Sauces should be prepared in advance.

2. Do not overcook hotbed chives and bean sprouts as both cook easily. Overcooking will result in sogginess.

醬瓜炒田雞

Frogs with Pickles

「雷峰夕照」是西湖標誌性的十景之一，古塔原為越王錢弘俶為慶祝寵妃黃妃誕子而建，所以又名黃妃塔。雷峰塔命途多舛，歷史上多次被戰火及人為破壞，原來的塔身在公元 1924 年倒塌。現在的雷峰塔是近年重建的，設計保留了宋代的風格，是中國第一座彩色銅雕寶塔。雷峰塔豎立西湖邊，風光綺麗，而「白蛇傳」中白娘子與許仙的淒美愛情故事，千百年來吸引遊人無數。

田雞又叫做水雞、蛤蟆，肉質鮮美嫩滑。杭州人喜歡吃蛤蟆（田雞），早在春秋時期已有記載。每年的端午節，杭州有給孩子吃蛤蟆的習俗，認為有下火的功效，夏天就不會生疔瘡。傳說「白蛇傳」中捧打鴛鴦的法海和尚，是蛤蟆精修煉成的，杭州人同情白娘子，所以要吃蛤蟆來懲罰他。

杭州醬瓜，又叫做脆瓜，台灣叫花瓜，味道鹹香帶甜，口感脆嫩，易於保存。做法是把黃瓜用鹽醃一天，用重物壓出瓜中水分，晾乾後放入醬油、辣椒、香料和糖煮成的醬汁中，浸兩三天就成了醬瓜，做法與醬蘿蔔相同。醬瓜以前是杭州家家戶戶都會做的漬物，現在超市有瓶裝出售。醬瓜是杭州人夏天餐桌上可口的小碟前菜，也可以用來做醬瓜炒毛豆、醬瓜炒杭椒、醬瓜炒肉，清代袁枚《隨園食單》中記載的「炒水雞」，即「醬瓜炒田雞」。

●材料

田雞 600 克	葱白 3 條	
鹽 1.5 茶匙	醬瓜 50 克	份量 4 人份

●做法

1. 田雞宰好去皮，把田雞的臂和腿切出成四件，田雞身不要，洗淨，用 1 茶匙鹽溶在 500 毫升水中，放入田雞浸泡 20 分鐘，撈出。

2. 田雞用水再沖洗一次，瀝乾水分，放在大碗中，加入生抽及薑汁拌勻醃 15 分鐘，再加入生粉拌勻。

3. 蒜頭切成兩段，葱白切段。

4. 中小火起油鑊把 500 毫升油燒至中溫（約 150℃），把田雞倒入用筷子弄散，泡油約 10 秒，即撈起瀝油，備用。

5. 把鑊裏的油倒起，只留 2 湯匙。大火燒熱鑊，先爆香蒜頭，再放入醬瓜炒勻。

6. 加入田雞，讚酒，放入蠔油、1/2 茶匙鹽、糖和葱白，炒至田雞全熟，加麻油兜勻，即成。

●烹調心得

1. 田雞在水中生長，水質如有污染，容易有寄生蟲，所以要用鹽水泡過。

2. 瓶裝的醬瓜在上海南貨店或台灣食品店有售。

Serves: 4 / **Soaking time:** 20 minutes / **Preparation time:** 20 minutes

Cooking time: 5 minutes

Ingredients

600 g frogs

1.5 tsp salt

1 tsp light soy sauce

1 tbsp ginger juice

1/2 tsp corn starch

6 cloves garlic

3 scallion stems

50 g pickles

1 tbsp Shaoxing wine

1 tbsp oyster sauce

1/2 tsp sugar

1/4 tsp sesame oil

Method

1. Peel frogs, separate arms and legs of each frog, discard the bodies, and rinse and soak in 500 ml water and 1 teaspoon of salt for 20 minutes. Drain.

2. Rinse frogs in fresh water, drain, and marinate with soy sauce and ginger juice for 15 minutes. Mix in corn starch.

3. Cut each clove of garlic into two pieces, and scallion stems into sections.

4. Heat 500 ml of oil to medium heat (about 150°C) put in frogs, and stir with chopsticks and deep fry over medium low heat for about 10 seconds. Remove frogs to drain.

5. Pour out oil leaving 2 tablespoons in the wok, stir fry garlic over high heat until pungent, and stir in pickles.

6. Put in frogs, drizzle wine, add oyster sauce, 1/2 teaspoon of salt, sugar and scallion stems, and stir fry until frogs are fully cooked. Add sesame oil, toss and transfer to plate.

Tips:

1. Soaking in salt water will help to clean the frogs.

2. Bottled pickles are available in stores selling Shanghai or Taiwan foods.

蟹炒年糕

Stir Fried Glutinous Rice Sticks with Crab

中菜的烹調技術，不斷發展了幾千年，與世界上大多數國家的烹調，有一個明顯的特色，就是烹調上的層次分明，使所有包括主材料、配料、調味料、香料等材料，經過不同的手法的預處理及烹調，達到菜式各種材料的最好效果，但最後又互相融合成為一個菜餚，而不只是裝盤上的主次配襯。

在烹飪的過程中，針對不同材料的性質，進行包括刀章、醃製、預處理、烹調的先後、火候、溫度等技術的配合，成為一道道截然不同的菜式。這些在中國人來說，看似理所當然亦習以為常的烹調，正正是中華民族幾千年來烹調智慧的積累，也是中國烹調追求完美境界的哲學。

杭州菜中的「蟹炒年糕」，就是一個好例子，材料中有形狀古怪而且有硬殼的蟹，有軟糯而容易黏鍋的江南水磨年糕，再加上一些會出水的紹菜絲，三種完全不同的材料，每種材料的形狀不同，軟硬不同，成熟時間不同，以至烹調手法完全不同，但最終炒成一碟美味的菜餚，而每種材料吃起來都是恰到好處，這就是中菜的烹調藝術，也體現了中華飲食文化的博大精深。

江南地區的年糕是由糯米（江米）磨粉做成，其中以浙江寧波的水磨年糕最為著名。有關年糕的文字記載，可追溯到一千多年前北魏的《齊民要術》，最早時是民間在農曆年夜時拜神祭祖的祭品，寓意五穀豐登、年年高升，所以稱為年糕。

份量　4 人份
準備時間　20 分鐘
烹調時間　10 分鐘

●材料

水磨年糕 300 克　　　　　海鮮醬 1 湯匙

　（大約是 6、7 條）　　　生抽 1/2 茶匙

蟹 600 克　　　　　　　　糖 1/2 茶匙

紹菜 100 克（切絲）　　　鹽 1/2 茶匙

薑片 10 克　　　　　　　　胡椒粉 1/4 茶匙

葱 2 根（切段）　　　　　麻油 1/2 茶匙

●做法

1. 年糕沖水洗淨，用刀斜切成半厘米厚片備用。

2. 螃蟹掀起蟹蓋，洗淨蟹身，再切成 6 到 8 塊，蟹鉗拍裂。

3. 在煲裏燒熱 2 公升水至出現蝦眼水（約 90℃），熄火，放下
　年糕，泡約 1-2 分鐘至稍軟，取出，用冷水泡涼，瀝乾，再
　用 1 湯匙油把年糕拌勻，使每一塊都沾上油。

4. 在鑊裏燒熱 500 毫升油至中低溫（約 130-140℃），放入年糕，
　用筷子把年糕分開，至稍為變黃，取出瀝油。

5. 放下螃蟹，轉中火炸至九成熟，取出瀝油。

6. 倒出鑊內的油，只留約 1 湯匙，大火把紹菜炒至軟身，取出。

7. 在鑊內燒熱 1 湯匙油，爆香薑葱，放入海鮮醬、生抽、糖和
　鹽炒香，加入螃蟹大火炒熟，再放入年糕和紹菜，兜勻，最
　後加胡椒粉和麻油即成。

●烹調心得

1. 蟹炒年糕用的蟹，可選用大閘蟹、毛蟹、膏蟹、紅蟳，
　以及奄仔蟹。

2. 水磨年糕是由糯米粉做成，加熱後很容易黏鍋。泡熱
　水時不要泡得太軟，取出後立即放入冷水，是不讓年
　糕的餘溫繼續把年糕煮得過軟。

Serves: 4 / Preparation time: 20 minutes / Cooking time: 10 minutes

Ingredients

300 g glutinous rice sticks
(about 6 to 7 rice sticks)

600 g crab

100 g Napa cabbage, shredded

10 g ginger slices

2 scallion, sectioned

1 tbsp hoisin sauce

1/2 tsp light soy sauce

1/2 tsp sugar

1/2 tsp salt

1/4 tsp ground white pepper

1/2 tsp sesame oil

Method

1. Rinse rick sticks, and slant cut into 1/2 cm thick slices.

2. Remove carapace, and clean and cut crab into 6 to 8 pieces. Crush the claws.

3. Heat 2 liters of water to just below boiling (about 90°C), turn off heat, put in rice sticks and soak for about 1-2 minutes until slightly soft. Remove rice sticks and rinse with fresh water until cooled. Drain and mix rice stick with 1 tablespoon of oil.

4. Heat 500 ml of oil in a wok to medium low heat (about 130-140°C), put in rice sticks and stir with chopsticks to gently deep fry until slightly yellow. Remove rice sticks to drain.

5. Deep fry crabs over medium heat until about 90% done, and remove to drain.

6. Pour out oil leaving about 1 tablespoon, and stir fry cabbage over high heat until soft. Remove cabbage.

7. Heat 1 tablespoon of oil in the wok and stir fry ginger and scallion until aromatic. Add hoisin sauce, soy sauce, sugar and salt, and put in crabs to stir fry over high heat until fully cooked. Put in rice sticks and cabbage and toss to mix. Add ground white pepper and sesame oil before transferring to plate.

Tips:

1. Both river crabs and salt water crabs can be used.

2. Glutinous rice sticks become soft and sticky when heated. Rinsing with fresh water immediately after soaking in hot water helps to prevent them from becoming too soft.

蟹釀橙

Stuffed Orange with Crab Meat

　　G20 是 Group of Twenty 的縮寫，是一個國際經濟合作論壇，於 1999 年 12 月 16 日在德國柏林成立，每兩年舉行一次，二十國集團的成員國 GDP 佔了全球 GDP 的 85%，人口佔全球人口的三分之二。2016 年的金秋九月，桂花飄香，在杭州舉行的 G20 第 11 次領導人峰會，吸引了全球的目光。中國為舉辦這次 G20 峰會做足準備，杭州市作為主辦城市，更是力求完美，絕對是傾城之作。

　　9 月 4 日晚，國家主席習近平夫婦在杭州西子賓館舉行國宴，招待各國領導人，菜式是費盡心思精選的杭幫菜。此次國宴菜單中，有一道不為外地人熟悉的「蟹釀橙」，引起了很多的注意和議論。一些對杭州菜不大認識的人，可能以為這是一道新派 Fusion 菜。其實，「蟹釀橙」是一道南宋時期的宮廷菜，根據宋末明初周密所著的雜記《武林舊事》，全書十卷，憶述南宋都城臨安（杭州）的歷史風貌，書中記載抗金名將張俊，以「蟹釀橙」進貢給宋高宗。

　　南宋文人林洪所撰寫的養生食療名著《山家清供》中，記載了「蟹釀橙」的做法：「橙大者，截頂，剜去穰，留少液，以蟹膏肉實其內。仍以枝頂覆之，入小甑，用酒、醋、水蒸熟。加苦酒入鹽供，既香而鮮。使人有新酒、菊花、香橙、螃蟹之興⋯」。自 G20 國宴之後，使更多人認識這道南宋名菜了。

份量　4 人份

準備時間　45 分鐘

烹調時間　30 分鐘

●材料

甜橙 4 個

蟹 600 克

麻油 1 湯匙

薑蓉 1 湯匙

花雕酒 6 湯匙

鎮江醋 2 湯匙

糖 1/2 湯匙

鹽 1/2 茶匙

生粉少許

做法

1. 將甜橙洗淨，在橙的圓底削去少許①～②，使橙可以站立不滾動。在橙上方用小刀刻出一個鋸齒形的蓋③～④。

2. 揭開橙蓋⑤～⑥，取出橙肉和橙汁，除去橙核和筋渣。橙肉切碎，和橙汁留用。

3. 把蟹蒸熟，剔取蟹肉和蟹黃成蟹粉，備用。

4. 燒熱炒鍋，加入麻油，燒至中低溫（約 120℃），爆香薑蓉，加入蟹粉稍為煸一下，放入橙肉及部份橙汁，加入 1/2 湯匙酒、1 湯匙醋、糖和鹽炒熱，用少許生粉勾芡，取出攤開至涼。

5. 分成 4 份，裝入橙中，蓋上橙蓋。

6. 把橙放在小碟中，調勻餘下的酒和醋，平均地放入碟中。

7. 用微波爐保鮮紙把橙連碟包住，用絲帶縈住，大火蒸 15 分鐘，取出。

8. 吃時解開絲帶，用匙羹直接食用即可。

①　②　③

④　⑤　⑥

●烹調心得

1. 挖橙肉時要留薄薄一層橙肉，避免蒸的時候橙皮的苦味滲入。

2. 放糖的份量多少，是要視乎橙的酸甜度而調整。

Serves: 4 / Preparation time: 45 minutes / Cooking time: 30 minutes

Ingredients

4 oranges

600 g crab

1 tbsp sesame oil

1 tbsp grated ginger

6 tbsp huadiao wine

2 tbsp Zhenjiang vinegar

1/2 tbsp sugar

1/2 tsp salt

corn starch

Method

1. Rinse and clean oranges. Cut the bottom off the oranges so that they can stand on their own ① ~ ② . With a small knife, carve a zigzag pattern near the top of the orange ③ ~ ④ .

2. Lift the cover of the oranges ⑤ ~ ⑥ , cut out the flesh but leave a thin layer of flesh on the rind. Remove pips and membrane holding the pulp, cut pulp into smaller pieces and save the juice.

3. Steam crab, and pick out meat and tomalley for later use.

4. Heat sesame oil in a wok to medium low heat (about 120°C) and stir fry ginger until aromatic. Put in crab meat and tomalley, and stir briefly. Add flesh and part of the juice from the oranges, and put in 1/2 tablespoon of wine, 1 tablespoon of vinegar, sugar and salt. Stir to mix until heated and thicken with corn starch. Remove to cool.

5. Divide filling into 4 portions and stuff each orange with 1 portion. Cover oranges with the orange tops.

6. Put oranges into 4 separate small plates. Mix the remaining wine and vinegar, and put into the plates.

7. Wrap each orange together with the plate with food grade cellophane wrap and tie with a ribbon. Steam over high heat for 15 minutes. Serve.

8. Untie ribbon and partake with a spoon directly.

Tips:

1. Leaving a thin layer of orange flesh on the rind will prevent the bitter flavor of the rind to seep into the stuffing when cooked.

2. The amount of sugar to add will depend of the sweetness of the oranges.

份量　6-8 人份

準備時間　15 分鐘

烹調時間　15 分鐘

浸漬時間　4 小時

酒香螺

Drunken Snails

　　在中國古代，人們吃螺的歷史悠久，特別是作為平民百姓的飯餐和小吃。品種方面，內地省份的人們主要是吃生長在湖泊、稻田和溝渠的淡水田螺和小螺螄；沿海省份則吃鹹淡水交界的泥螺，以及各式大小海螺，包括大海螺、紅螺、花螺（方斑東風螺）、東風螺（泥東風螺）等。

　　古人相信吃螺有食療作用，有清熱利水，除濕解毒的功效。《本草拾遺》中關於吃螺食療的記載：「煮食之，利大小便，去腹中結熱、目下黃、腳氣沖上、小腹結硬、小便赤澀、腳手浮腫；生浸取汁飲之，止消渴；碎其肉敷熱瘡。」

　　中國圓田螺，又稱為香螺。每年秋天，江南地區的田螺是人們喜愛的流行時節食品。「酒香螺」是南宋時期，臨安（杭州）的市井名菜，根據宋代吳自牧所撰《夢粱錄》記載，當時的「酒香螺」用的材料是田螺。「酒香螺」，即醉螺，我們以古代食譜為基礎，再作改良，向讀者介紹這道美味的古代杭州民間菜。

　　材料方面，可選用田螺、東風螺或花螺。由於市場上田螺並不一定常見，所以食譜是改以花螺為材料。為了更適合現代人對食物衛生的要求，我們將原來的生醉，改為熟醉，以及用玻璃瓶或密實食物袋代替古代的泥罈。

材料

活花螺（有殼）900 克	紹興酒 100 毫升
鹽 2 茶匙	薑絲 40 克
花椒粒 1 湯匙	芫荽 2 株（切碎）
碎冰糖 25 克	

活花螺

做法

1. 大火煮沸大鍋水，把花螺倒入，見鍋中的水再煮至沸騰，沸煮 1 分鐘後立即撈出花螺。放涼後挑出螺肉（約得 300 克），撕去內臟，再用水洗淨。

2. 把 1 茶匙鹽放入 500 毫升冷開水中拌至溶，放入花螺肉浸泡 15 分鐘，取出，用乾淨棉布吸乾水分。

3. 用 50 毫升清水，加入花椒粒，煮沸後轉小火煮 5 分鐘，用漏網隔去花椒粒不要，取花椒水放涼，備用。

4. 用 150 毫升清水，放入冰糖煮溶，放至涼，拌入紹興酒、薑絲、花椒水、1 茶匙鹽拌勻成醉汁，倒入一個乾淨玻璃瓶中或密實食物袋中。

5. 加入花螺肉，醉汁要浸過螺肉。密封，放在冰箱中約 4 至 5 小時。

6. 取出花螺肉盛碟，灑上芫荽碎，即成。

●烹調心得

1. 灼花螺要注意時間，不能煮過火，否則螺肉會韌。

2. 花螺用醉汁浸漬半天已可食，但不宜超過 1 天才進食。

3. 也可以用東風螺，灼的時間要稍加一兩分鐘。

Serves: 6-8 / Preparation time: 15 minutes / Cooking time: 15 minutes

Soaking time: 4 hours

Ingredients

900 g Areola Babylon

2 tsp salt

1 tbsp Sichuan peppers

25 g crushed rock sugar

100 ml Shaoxing wine

40 g shredded ginger

2 coriander, chopped

Method

1. Bring to a boil a large pot of water, put in snails, re-boil for 1 minute, drain and allow to cool. Pick the flesh (about 300 g) out of the shells, remove organs, and rinse with fresh water.

2. Dissolve 1 teaspoon of salt in 500 ml of cold drinking water and soak snail flesh for 15 minutes. Drain and pat dry with clean towels.

3. Boil Sichuan peppers in 50 ml of water over low heat for 5 minutes. Run and save water through a sieve and allow to cool. Discard Sichuan peppers.

4. Heat 150 ml of water and dissolve rock sugar. Allow to cool and add wine, ginger, Sichuan pepper water and 1 teaspoon of salt into a wine sauce. Pour into a clean glass jar or a food storage bag.

5. Add the snail flesh, seal and refrigerate for 4 to 5 hours. The flesh must be totally covered by the wine sauce.

6. Transfer the snails to a plate and top with chopped coriander.

Tips:

1. Don't overcook the snails or the flesh will become too chewy.

2. Soak snails for a minimum of 4 hours but not over 24 hours.

3. Other snails such as babylonia lutosa can also be used but may need to cook additional 1 to 2 minutes.

紅燉蘿蔔

Braised Turnips

份量 4 人份

準備時間 10 分鐘

烹調時間 1 小時 30 分鐘

　　杭州菜和潮州菜都有一種相同的特色烹調方法，就是「素菜葷做」，主角是素菜，但是在烹調過程中大用葷料，取其肉味和油脂，上桌前一定把葷料盡棄，看起來是素菜，但它又絕不是齋菜，是介乎葷素之間的菜式。這種做法，在各省中菜烹調中很少見，它別出心裁的做法，正好表達了「有味者使之出，無味者使之入」的烹調境界，足見杭州和潮州兩地，皆有深厚的文化底蘊。

　　「素菜葷做」的菜式，源自官府菜，平民百姓是不會捨得把同煮的肉棄掉而只取其味，只有官府和富貴人家的私房廚師才會這樣做，既以美味的素菜討好主人，而下人也有大肉吃，一舉兩得，但又深藏不露。潮州菜的「素菜葷做」代表菜式是家傳戶曉的「厚菇大芥菜」，而杭州菜中最具代表性的菜式是「紅燉蘿蔔」。

　　白蘿蔔是營養豐富的根莖類蔬菜，常用於涼拌、燜煮、燉湯、醬漬、醋漬、泡菜等菜式，具健胃消食、化痰生津和利尿的食療功效。浙江杭州南面的蕭山，地處錢塘江南岸，是杭州蕭山國際機場的所在地。蕭山的著名特產是白蘿蔔，味甜無渣，水透靈瓏，是烹調杭州家常菜「紅燉蘿蔔」的最佳選擇。

●材料

白蘿蔔 750 克	葱 2 根（切段）
鹽 1 茶匙	清雞湯 500 毫升
紅麴米 1 湯匙	老抽 1 茶匙
五花腩 200 克（切片）	冰糖 10 克
薑片 20 克	蠔油 1 湯匙

●做法

1. 白蘿蔔削皮，用滾刀切成大塊，放在鍋內，加冷水至完全覆蓋蘿蔔，加鹽，煮沸後，轉小火煮 30 分鐘。

2. 取出蘿蔔，用冷水沖洗，瀝乾水分。

3. 紅麴米用香料袋或茶袋裝好，備用。

4. 在乾淨鍋內燒熱 2 湯匙油，煸炒五花腩約 1 分鐘，加薑蔥爆香。

5. 放入雞湯、蘿蔔、紅麴米、老抽、冰糖和蠔油，大火沸煮 10 分鐘，轉中小火再煮 45 分鐘，取出蘿蔔盛碟，五花腩和薑蔥都不要。

6. 把鍋裏的湯汁用大火收至稍稠，淋在蘿蔔上，即成。

●烹調心得

紅麴米天然的紅色，用來增加蘿蔔的顏色，對味道沒有影響；可以不用，也可以在煮至顏色夠紅時取出。

Serves: 4 / **Preparation time:** 10 minutes / **Cooking time:** 1 hour 30 minutes

Ingredients

750 g turnips

1 tsp salt

1 tbsp red yeast rice

200 g pork belly slices

20 g ginger slices

2 scallion, sectioned

500 ml chicken broth

1 tsp dark soy sauce

10 g rock sugar

1 tbsp oyster sauce

Method

1. Peel turnip and roll cut into large pieces. Put turnip into a wok, add water to cover turnip, add salt, bring to a boil, and reduce to low heat to cook for another 30 minutes.

2. Rinse turnip with fresh water and drain.

3. Put red yeast rice into a spice bag.

4. Heat 2 tablespoons of oil in a wok and stir fry pork belly for about 1 minute. Add ginger and scallion.

5. Put in chicken broth, turnip, red yeast rice, soy sauce, rock sugar and oyster sauce, boil over high heat for 10 minutes, and reduce to medium low heat and cook for another 45 minutes. Discard pork belly, ginger and scallion, and remove turnip to a deep plate.

6. Reduce sauce in the wok and drizzle on the turnip.

Tips:

Red yeast rice is a natural color agent and can be taken out whenever the desired color is obtained. It does not affect the flavor of the food.

火腿冬瓜夾

$$\diamond$$

Winter Melon with Ham

　　這是一道夏令的宴客菜式，清代著名文學家袁枚所著《隨園食單》中曰：「冬瓜之用之最，拌燕窩、魚、肉、鱔、火腿皆可。」冬瓜味道雖然寡淡，但配上鹹香的金華火腿片，葷素搭配，相得益彰，成為一道清麗不俗的菜式。

　　冬瓜是瓜果蔬菜中，唯一不含脂肪的食物。中醫認為冬瓜有清熱解暑、潤肺消痰、利小便、治腳氣等功效。冬瓜含丙醇二酸成份，能抑制糖類物質轉化為脂肪，所以有減肥瘦身的功效。冬瓜含豐富維生素Ｃ，鈉鹽含量低，對患有高血壓、腎病、糖尿病、浮腫等患者，冬瓜是最理想的食物之一。

　　「火腿冬瓜夾」這道菜，已經有起碼百多年歷史，出處不詳，浙江金華的菜式中有一道「火腿冬瓜方」，但做法並不完全相同，是加了蝦米再用酒烹。據先父特級校對在《食經》中記載「火腿冬瓜夾」的古老做法，是把帶肥的金華火腿切片，放入冬瓜片縫中夾好，放在鍋中油煎，把冬瓜煎至微黃，而火腿油也溢出，以碟盛起，加以上湯再蒸，讓冬瓜吸收上湯和火腿的味道，冬瓜酥嫩，火腿鹹香。

　　今天人們重視健康，我們覺得這道菜應該做得更清淡，所以稍為改變了父親的老做法。先把火腿片蒸過，棄用蒸出來的油和水，把火腿片夾在冬瓜片中，也不用油煎，而是直接加入上湯或雞湯浸着冬瓜，放在鍋中蒸至熟，這樣做減少了大部份火腿的鹹味和油份，吃得健康，也更矜貴。

材料

冬瓜 600 克　　　　　　薑 3 片

火腿 100 克　　　　　　雞湯 250 毫升

蜂蜜 1 湯匙

做法

1. 削去冬瓜皮，但不要削得太深，留少許瓜青色 ①。

2. 把冬瓜切成約 4 厘米厚的四塊 ②，去掉瓜瓤 ③。

3. 把每一塊冬瓜切為四件，每一件再切一刀至瓜青，但不要切斷 ④，即每一塊冬瓜中間都有一條不切斷的刀縫可夾火腿 ⑤。

4. 火腿切薄片，加蜂蜜拌勻 ⑥，大火蒸 3 分鐘。

5. 在冬瓜中間的刀縫中，插入一片薄火腿 ⑦，排好在蒸碟中，中央可放一些火腿絲做裝飾。

6. 冬瓜夾上面放上薑片，倒入雞湯 ⑧浸着冬瓜，大火蒸約 12 分鐘，挾出薑片棄掉，即可上桌。

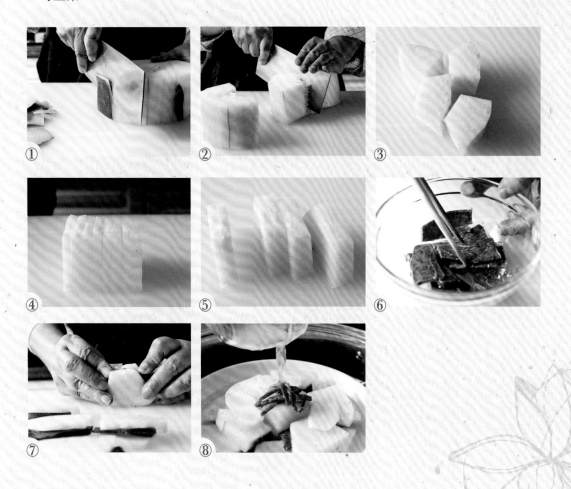

Serves: 6-8 / Preparation time: 20 minutes / Cooking time: 12 minutes

Ingredients

600 g winter melon

100 g Jinhua ham

1 tbsp honey

3 ginger slices

250 ml chicken broth

Method

1. Peel winter melon but leave a trace of green on the surface ① .

2. Cut winter melon into 4 equal size chunks each about 4 cm thick ② . Remove rinds and piths ③ .

3. For each chunk of melon, cut into 4 equal pieces, and make a deep cut on each piece but without cutting through ④ to allow for insertion of a piece of ham ⑤ .

4. Cut ham into slices, mix with honey ⑥ and steam for 3 minutes.

5. Insert a slice of ham into each piece of melon ⑦ , and arrange melon pieces on a plate.

6. Top with ginger, add chicken broth ⑧ and steam for 12 minutes. Discard ginger before serving.

份量 4-6 人份
準備時間 20 分鐘

脆 琅 玕（薑汁萵苣）

Celtuce Salad

琅玕（粵音 longgon 浪干），意思是指美玉。魏國曹操之子，才子曹植的《美女篇》中詩云：「美女妖且閒，採桑歧路間。柔條紛冉冉，落葉何翩翩。攘袖見素手，皓腕約金環。頭上金爵釵，腰佩翠琅玕。」

「脆琅玕」是一道南宋的杭州開胃涼菜，型如美玉，爽脆清甜。南宋文學家林洪，在其著作《山家清供》中記載這道美食：「萵苣去葉皮，寸切，搗薑、鹽、熟油、醋拌漬之，頗甘脆。杜甫種此，二旬不甲坼，且嘆君晚得微祿，坎坷不進，猶芝蘭困荊杞。以是知詩人非為口腹之奉，實有感而作也。」意思是說著唐朝名詩人杜甫在他仕途不得志時，辟菜園種植萵苣，他一連二十天都不願離開菜園，而且老在嘆息俸祿微薄，感到自己前途坎坷，就像芝蘭被困於荊杞叢中；由此可見，他流連菜園不歸，並不是為了口腹（吃萵苣），而是有感而為。

杜甫在《玄都壇歌寄元逸人》中詩句曰：「知君此計成長往，芝草琅玕日應長。」指琅玕（萵苣）為仙家飲食，長生靈藥。萵苣，又名萵筍、苣筍，原產於地中海沿岸，大約在漢代時期傳入中國。萵苣含有大量的萵苣素和胡蘿蔔素，以及維生素 B_1、B_2 和 C_4。萵苣可生食涼拌，亦可作炒菜，吃萵苣時，大多數人只吃莖肉，因莖部肉質脆嫩，味道清甜；但其實萵苣的葉子更具營養價值，卻因其口感粗硬，中國人一般不會用來做菜餚。但在西方近年流行吃粗纖維的植物，也有把萵苣葉煮熟，加鹽和橄欖油來做拌食菜。

🍃材料

萵苣 2 根	鎮江醋 2 湯匙
鹽 1 茶匙	薑汁 3 湯匙
糖 2 湯匙	麻油 1 湯匙

萵苣

●做法

1. 萵苣去葉只留莖部，切掉老的部份，削去外皮，露出萵苣心，再切成約 1/2 厘米厚 4x2 厘米的萵苣塊。①～⑨

2. 把萵苣放在笪箕內，用 1/2 茶匙鹽醃 10 分鐘，用廚紙吸乾，排好在盤中。⑩～⑪

3. 把 1/2 茶匙鹽、糖、醋、薑汁和麻油拌勻成醬汁，淋在萵苣上，即成。

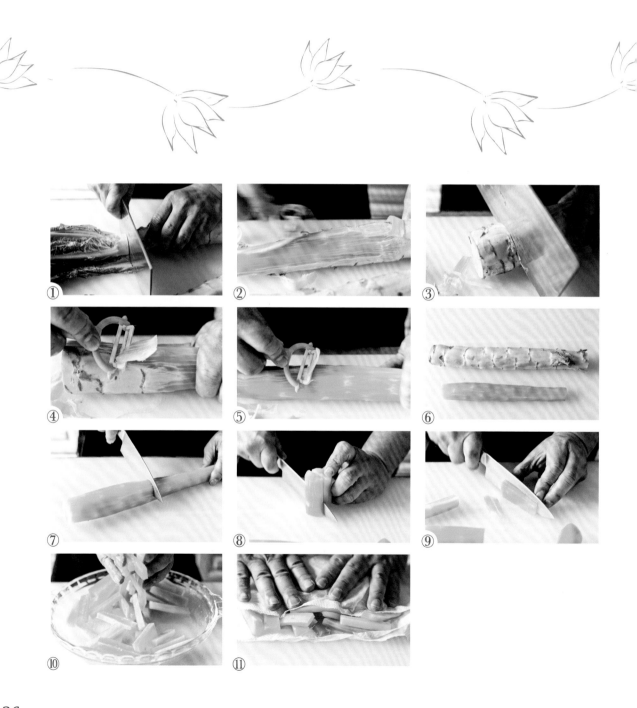

Serves: 4-6 / Preparation time: 20 minutes

Ingredients

2 celtuce

1 tsp salt

2 tbsp sugar

2 tbsp Zhenjiang vinegar

3 tbsp ginger juice

1 tbsp sesame oil

Method

1. Remove the leaves, peel the celtuce stems, keeping only the tender heart and cut into 1/2 cm thick, 4 x 2 cm pieces. ① ~ ⑨

2. Place celtuce pieces in a colander and marinate with 1/2 teaspoon of salt for 10 minutes. Pat dry with kitchen towels and arrange on a plate. ⑩ ~ ⑪

3. Mix 1/2 teaspoon of salt, sugar, vinegar, ginger juice and sesame oil into a salad dressing and drizzle on the celtuce.

「南宋杭幫菜」脆琅玕（薑汁萵苣）

葱油蘿蔔絲

Shredded Turnip with Scallion Flavored Oil

　　愛吃上海菜和小食的朋友，一定吃過上海的「葱油拌麵」，葱油＋醬油＋麵，就成了香噴噴的麵食。「葱油拌麵」的靈魂來自葱油，拌麵的葱油要做得好，要耐心地用文火慢熬，把葱的味道用油迫出來，這樣做的葱油才會香噴噴，心急則不行。學會了做葱油，有空就熬一瓶在家，炒菜炒飯拌麵隨時可用，方便得很。

　　明明只是一道清淡的素涼菜，對於杭州人來說，也可以吃得千姿百態。明代李時珍對蘿蔔有如此的評價：「可生可熟，可菹可醬，可豉可醋，可糖可臘可飯，乃蔬中之最有益者。」「葱油蘿蔔絲」，是一道做法簡單的涼拌菜，我們套用了上海「葱油拌麵」的葱油製法，更覺倍添江南風味。在白蘿蔔當造的秋冬季節，買一個白蘿蔔和一束小葱，熬好了葱油，把白蘿蔔刨成絲，就可以做出清爽可口的「葱油蘿蔔絲」。

材料

白蘿蔔 600 克
鹽 1 茶匙
胡椒粉 1/4 茶匙
葱 8-10 根
葱花 2 湯匙

做法

1. 蘿蔔削皮後刨絲，放在笪箕內，用鹽和胡椒粉拌勻。
2. 葱切葱段，用 6 湯匙油慢火把葱段煎至變黑色，取出葱段。
3. 把蘿蔔絲裝盤，撒上葱花，放上幾根煎過的葱段，淋上燒熱的葱油即成。
4. 吃時把蘿蔔絲和葱油拌勻。

份量 4-6 人份
準備時間 15 分鐘
烹調時間 15 分鐘

Serves: 4-6 / Preparation time: 15 minutes / Cooking time: 15 minutes

Ingredients

600 g turnip
1 tsp salt
1/4 tsp ground white pepper
8-10 scallion
2 tbsp chopped scallion

Method

1. Peel and shred turnip, mix with salt and ground white pepper, and put into a colander.
2. Cut scallion into sections, and pan fry slowly in 6 tablespoons of oil until turning a dark color and the oil takes on a scallion flavor. Take out scallion.
3. Transfer turnips to a plate, top with chopped scallion and a few pieces of pan fried scallion, and drizzle heated flavored oil.
4. Mix shredded turnip with the flavored oil at the table.

醬烤花菜

Sti Fried Cauliflower in Sauce

　　花菜，又名椰菜花，是中菜和西菜常用的蔬菜。花菜有很多不同的品種，有白色、紫色和青色，而白色的花菜在市場上最為常見。花菜總是那麼平平無奇，沒有特殊的滋味，菜式上很難獨當一面，很少人對它特別寵愛，也不會覺得討厭，反正一年四季花菜都會靜悄悄地躺在市場的菜架上。

　　近年市場上出現一種叫福建花菜的品種，白色的菜花，粉綠色的莖，就像淡掃娥眉的小家碧玉，穿着粉綠的衣裳，眼波流動，不禁令人注目。這種花菜由福建運來，商販們不知其名，只好叫它做福建花菜。這種花菜的特點是花頭比較散開，讓人一眼就看到它那粉綠色的莖。自從吃過這種福建花菜，我們就貪新忘舊，喜歡上了它。福建花菜除了顏色漂亮之外，味道清甜，口感爽脆，用來做「醬烤花菜」最為適合，就讓花菜當一回主角吧！

● 材料

福建花菜 600 克	海鮮醬 2 湯匙
鹽 1 茶匙	蠔油 2 湯匙
蒜頭 3 瓣	

份量　4 人份
準備時間　10 分鐘
烹調時間　5 分鐘

◎ 做法

1. 福建花菜只要菜花部份，硬的菜梗要切掉。

2. 把鹽加入 500 毫升清水中，放入福建花菜泡 15 分鐘取出；另煮沸一鍋水，把花菜汆燙 1 分半鐘，瀝乾水分。

3. 蒜頭去衣，每瓣切成兩粒，備用。

4. 燒熱 2 湯匙油，用中火爆香蒜粒，加海鮮醬和蠔油炒勻。放入花菜，轉大火爆炒約 3 分鐘，即成。

● Ingredients

600 g Fujian cauliflower

1 tsp salt

3 cloves garlic

2 tbsp hoisin sauce

2 tbsp oyster sauce

● Method

1. Discard the stems of the cauliflowers, keeping only the blossoms and cut into smaller pieces.

2. Soak blossoms in 500 ml of water with salt for 15 minutes, drain, and blanch blossoms for one and half minutes. Drain.

3. Peel and cut each clove garlic into two halves.

4. Heat 2 tablespoons of oil, and stir fry garlic over medium heat until aromatic. Stir in hoisin sauce and oyster sauce, add blossoms and stir fry over high heat for about 3 minutes. Transfer to plate.

糟 油 茭 白

Sautéed Water Bamboo with Fermented Glutinous Rice

杭州、紹興和太湖地區盛產茭白，它與蓴菜、鱸魚合稱江南三大名產。茭白又名茭瓜，古代稱為「菰米」。廣東人稱茭白為茭筍，只因茭白尖長的形狀似竹筍，但茭白與竹筍毫無關係，它是禾本科水生植物，生長在池塘和湖泊地區。茭白非但不熱毒，還是解毒之物，能清暑止渴、利尿，更有解酒清熱的功效。茭白含豐富的蛋白質和維生素 A，茭白的纖維更有助腸道蠕動，改善便秘，而且含熱量低，是一種健康蔬菜。

清代袁枚《隨園食單》中介紹茭白：「炒肉，炒雞俱可，切整段，醬醋炙之，尤佳。煨肉亦佳，須切片，以寸為度。」紹興民間有更簡單的吃茭白方法，叫做「飯焐茭白」，就是把茭白去衣剝好，放在飯上同蒸，吃時蘸醬油，滋味無窮。

「糟油茭白」是一道傳統的素菜。杭州和紹興地區的酒糟，是在用糯米釀製黃酒剩下的酒渣，密封半年以上，即為香糟，酒味濃郁。酒糟主要有三種，還有福建菜和台灣菜用的是紅糟，和山東菜用的黃糟，由於各地釀酒的原料和工藝不同，製出來的酒糟也不同。但是，由於酒糟畢竟是加工後的渣滓，衛生質素難以控制，因此近年就算是當地的食肆，也少用釀酒後的酒糟做菜，取而代之的是酒釀，以蒸熟的糯米加菌種發酵而成，做法簡單，不少人是自家製造。酒釀也叫做醪糟，我們在超市或南貨見到盒裝或瓶裝、白色的甜酒浸着一些糯米粒的就是酒釀。

我家做「糟油茭白」，配上一條京葱，京葱也稱為大葱、北葱、韭葱，味道芳香而帶甜，加在清淡孤寡的茭白中，頓使味道變得更有層次，生色不少。

份量 4-6 人份

準備時間 10 分鐘

烹調時間 5 分鐘

茭白

● 材料

茭白 450 克

京葱 1 根

酒釀 50 毫升

鹽 1/2 茶匙

薑 10 克（切絲）

麻油 1 茶匙

● 做法

1. 茭白切去綠色部份，刨淨白色的茭白肉，用滾刀切成長塊。

2. 煮大鍋水，把茭白放入煮約 2 分鐘，取出，瀝乾水分。

3. 京葱斜切成片，備用。

4. 酒釀加鹽倒入攪拌機中打成香糟汁。

5. 大火燒熱 3 湯匙油，爆香薑絲，放入京葱同炒約 1 分鐘，加入茭白塊炒 1 分鐘，加入香糟汁煮沸炒勻，加入麻油兜勻即成。

● 烹調心得

1. 酒釀在超市或南貨店有售。

2. 茭白是水生植物，為怕湖水有污染，所以要用水汆煮過才炒，更不宜生吃。

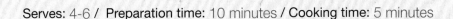

Serves: 4-6 / **Preparation time:** 10 minutes / **Cooking time:** 5 minutes

Ingredients

50 g water bamboo

1 Beijing scallion

50 ml fermented glutinous rice

1/2 tsp salt

10 g shredded ginger

1 tsp sesame oil

Method

1. Peel water bamboo to expose the white core, and roll cut into pieces.

2. Blanch water bamboo for about 2 minutes.

3. Slant cut Beijing scallion.

4. Blend fermented glutinous rice together with salt into a fermented glutinous rice sauce.

5. Heat 3 tablespoons of oil in a wok, stir fry ginger until fragrant, add Beijing scallion and stir for 1 minute. Put in water bamboo and stir fry for 1 minute, add fermented glutinous rice sauce and toss until the sauce is reduced. Stir in sesame oil and transfer to plate.

Tips:

1. White fermented glutinous rice is available in supermarkets and Shanghainese food stores.

2. Water bamboo should always be cooked and not eaten raw.

份量 4-6 人份

準備時間 15 分鐘

烹調時間 10 分鐘

炒豆腐鬆

Stir Fried Diced Tofu

　　有關豆腐的發明者，眾說紛紜，有說是西元前 164 年，淮南王劉安在八公山上採藥煉丹時，無意中以石膏點豆汁得來的靈感而發明的；也有考證說豆腐是在唐代或五代才有文獻的記載，所以應該把豆腐的發明年代推遲到唐末五代。可喜的是，在 1963 年被公佈為河南省重點保護文物單位的河南新密市打虎亭東漢墓裏出土的石刻畫中有一幅《製作豆腐工藝圖》，把製作豆腐的年代推前到起碼 1800 年前的東漢，也就是劉安後三百多年，比唐代早四百多年。由此可見，劉安是豆腐發明者的說法是很有可能的。豆腐性涼，有清熱、生津潤燥的作用。豆腐價格平宜，烹調容易，是最受歡迎的食材之一。

　　著名的山東孔府菜，有一道名菜是「炒豆腐泥」，其做法與浙江的「炒豆腐鬆」有些相似，但豆腐是壓碎的，材料中加了雪裡蕻（雪菜）、酒和薑，其他配料比浙江的「炒豆腐鬆」為少。浙江傳統的「炒豆腐鬆」更具江南風味，豆腐是切粒的，配料有蝦米、冬菇、火腿、雞肉或豬肉粒。其實家常菜就是家中有什麼就隨手放什麼，各家各法，沒有所謂正宗不正宗。這道菜簡單易做，是一道美味可口、老少咸宜的菜式。

●材料

硬豆腐 1 塊 (約 250 克)　　雪裡蕻 75 克

鹽 1 茶匙 (泡豆腐用)　　薑米 1 湯匙

蝦米 15 克　　絞豬肉 50 克

冬菇 2-3 朵　　鹽約 1/4 茶匙

金華火腿 10 克

●做法

1. 豆腐切成 1 厘米小丁，煮沸 1 公升清水，放 1 茶匙鹽煮溶，放入豆腐，熄火，浸 2 分鐘，撈起瀝水。

2. 蝦米浸軟，切小粒。

3. 冬菇浸透，去蒂切成約 0.5 厘米小粒。

4. 金華火腿蒸 5 分鐘，取出切成約 0.3 厘米小粒。

5. 雪裡蕻切去菜頭和葉不要，把莖洗淨，浸水 10 分鐘，瀝乾水分，切成 0.5 厘米小粒。

6. 燒熱 3 湯匙油，下薑米爆香，放入冬菇和絞豬肉同炒至熟。

7. 加入蝦米、雪裡蕻一起炒勻，放入豆腐、火腿和鹽，同炒約 3 分鐘即成。

●烹調心得

1. 硬豆腐即適合煎炸用的豆腐。

2. 市場上的雪裡蕻有兩種，鮮綠色的和黃綠色的，這道菜適合選用鮮綠色的。

3. 這一道菜的鹹味主要來自雪裡蕻、蝦米和火腿，在加鹽調味前要先試味。

Serves: 4-6

Preparation time: 15 minutes

Cooking time: 10 minutes

Ingredients

1 firm tofu (about 250 g)

1 tsp salt (for soaking tofu)

15 g dried shrimps

2-3 dried black mushrooms

10 g Jinhua ham

75 g pickled potherb mustard

1 tbsp ginger,, chopped

50 g minced pork

1/4 tsp salt

Method

1. Dice tofu into 1 cm cubes. Dissolve 1 teaspoon of salt in 1 liter of boiling water, put in tofu, turn off the heat and soak for 2 minutes. Drain.

2. Soften dried shrimps with water and cut into small pieces.

3. Soften mushrooms in water, remove stems, and cut into 0.5 cm pieces.

4. Steam Jinhua ham for 5 minutes and dice into 0.3 cm cubes.

5. Remove the leaves and the head of the potherb mustard, keeping only the stems. Rinse, soak in water for 10 minutes, drain and cut the stems into 0.5 cm pieces.

6. Heat 3 tablespoons of oil in a wok, stir fry ginger, add mushrooms and minced pork and stir fry until fully cooked.

7. Stir in dried shrimps and potherb mustard, put in tofu, ham and salt and stir fry for about 3 minutes. Transfer to plate.

Tips:

1. Firm tofu is the kind often used for pan frying or deep frying.

2. Two kinds of pickled potherb mustard, bright green and yellowish green, are available. The bright green kind should be used.

3. Most of the salty flavor comes from pickled potherb, dried shrimps and ham. Tasting is advised before adding salt.

「浙江農家菜」炒豆腐鬆

份量　4-6 人份
準備時間　15 分鐘
烹調時間　15 分鐘

油燜春筍

Sautéed Spring Bamboo Shoots

　　竹是溫帶和熱帶植物，中國是世界上產竹最多的國家，擁有超過一百多個品種，分佈在全國各地。竹筍是竹的幼芽，顧名思義，春天破土而出的是「春筍」；冬季收藏在土中的便是「冬筍」。竹筍性寒味甘，有滋陰涼血、清熱化痰、利尿通便的功效。

　　杭州人吃筍，吃法甚多，有涼拌筍、油烤筍、醬燒筍、炒雪筍、蝦子春筍、青筍炒肉絲、炒筍衣，和名菜南肉春筍等等。杭州人很固執，吃筍是不時不食，不鮮不食，春天吃春筍，夏秋季節吃鞭筍，深秋入冬吃冬筍。在杭州天目山地區出產的春筍，嫩滑無渣，爽脆鮮甜，是春筍中之極品。

　　鮮嫩爽甜的「油燜春筍」，把春筍炒得原汁原味的是杭州菜，炒得濃油赤醬的是上海菜。春筍剝去了大部份的外殼，剩下的嫩白如玉的筍尖，是這道菜唯一的材料，其實成本頗高。

　　「油燜春筍」一般有兩種做法，一種是先炸再煸炒，然後加水和醬料燜至汁稠，一般餐館廚師會用這種做法；第二種是生炒，用油把筍炒至熟，加醬油和糖等調味料再炒至收汁，中途不加水不加蓋。以下介紹的是比較清爽的第二種做法，感謝杭州菜名店西湖春天董事長張曉光先生賜教。

●材料

鮮嫩春筍（連殼）1 千克

老抽 1.5 湯匙

鹽 1 茶匙

紅糖 1 茶匙

麻油 1 茶匙

●烹調心得

1. 春筍最脆嫩的季節很短，過了農曆三月的筍就會開始變老，亦會有澀味，只適合製造筍乾。

2. 用紅糖來炒春筍，增加焦糖的顏色和味道，效果更佳。

●做法

1. 用刀沿着春筍的長度把筍衣刈開，把筍衣一層一層地剝開
   到筍芯。把筍底部纖維較硬的部份切除。① ~ ⑪

2. 用刀把春筍拍裂或切成條，再把每一條折斷成兩段。

3. 燒熱 3 湯匙匙油，把筍段放入鑊中用中小火炒 10 分鐘。

4. 放入老抽、鹽、紅糖，再繼續不斷炒 5 分鐘。

5. 大火收汁，加麻油，炒勻即成。

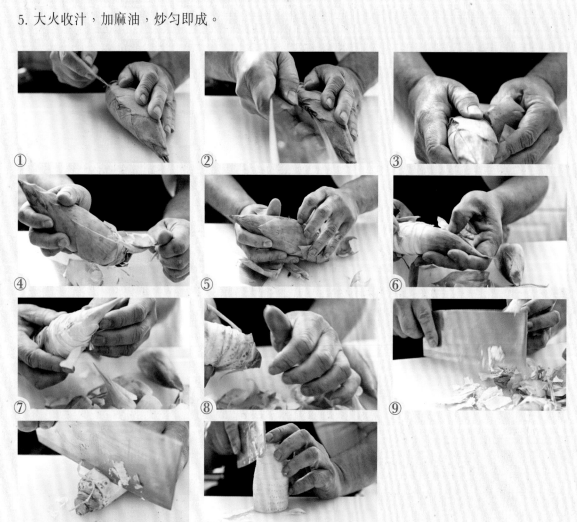

① ② ③
④ ⑤ ⑥
⑦ ⑧ ⑨
⑩ ⑪

Serves: 4-6 / Preparation time: 15 minutes / Cooking time: 15 minutes

Ingredients

1 kg fresh spring bamboo shoots

1.5 tbsp dark soy sauce

1 tsp salt

1 tsp red sugar

1 tsp sesame oil

Tips:

1. Spring bamboo shoots are best before the end of March of the Chinese lunar calendar, after which time they will become fiberous and coarse.

2. Red sugar enhances the color of the bamboo shoots as well as giving them a caramelized flavor.

Method

1. Slit open the skin of the shoots along its length and peel to expose the heart. Cut off the bottom of the shoot where the fibres are tough. ① ~ ⑪

2. Cut or smash the heart of each shoot into thick strips, and break each into two lengths.

3. Heat 3 tablespoons of oil in a wok, add bamboo shoots and sauté over medium low heat for about 10 minutes.

4. Add soy sauce, salt and sugar, and sauté for another 5 minutes.

5. Reduce over high heat, and stir in sesame oil.

杭州西湖沿岸的山坡上，種了很多桂花樹，屬木犀科的大樹，生長於江南一帶，四季常綠，枝葉茂盛，開花時芳香四溢，所以桂花樹自古多作為大宅庭園的觀賞植物，唐末李商隱詩云：「昨夜西施涼滿露，桂花吹斷月中香」。桂花樹在農曆八月以至整個秋冬都會開花，一束束金黃色小花，掛滿樹枝，而且花期較長，不易脫落。桂花味甘性溫和，有暖胃平肝、活絡化痰的功效。

桂花白玉是一道江南甜食，亦是一道可口的素涼菜。白玉是新鮮淮山（山藥），淮山本名懷山藥，最早原產於河南省懷慶的懷山。中醫認為淮山有安神、補肺腎、益胃健脾、助五臟、強筋骨的功效，主治脾胃虛弱、食慾不振、腰酸背痛、易疲倦、肺虛心燥、痰喘咳嗽、尿頻、高血脂和肥胖等病症。多吃淮山有益健康，延年益壽，是食療佳品。

● 材料

新鮮淮山 500 克
白糖 150 克
桂花糖 2 湯匙
乾桂花 2 湯匙

● 做法

1. 把白糖融化在 750 毫升水內，再放入冰箱內至涼。
2. 新鮮淮山削去皮，斜切成約 1/2 厘米厚片，蒸 12-15 分鐘。
3. 把淮山放入冰凍的糖水內泡 2 小時。取出瀝乾。
4. 拌入桂花糖，裝盤後撒上乾桂花即成。

份量　前菜小碟
準備時間　15 分鐘
烹調時間　15 分鐘
浸泡時間　2 小時

Serves: appetizer / Preparation time: 15 minutes
Cooking time: 15 minutes / Soaking time: 2 hours

● Ingredients

500 g fresh Chinese yam
150 g sugar
2 tbsp osmanthus sugar
2 tbsp dried osmanthus

● Method

1. Dissolve sugar in 750 ml of water to make a thin syrup and refrigerate.
2. Peel Chinese yams and slant cut into 1/2 cm thick slices. Steam for 12-15 minutes.
3. Soak Chinese yam in syrup and refrigerate for 2 hours, drain.
4. Mix in osmanthus sugar, transfer to plate and top with dried osmanthus flower.

桂 花 白 玉

◇

Chinese Yam with Osmanthus

份量 4-6 人份
準備時間 10 分鐘
烹調時間 10 分鐘

蟹粉豆腐羹

Tofu Soup with Tomalley and Crab Meat

　　江河湖海都產螃蟹，但江浙人都認為海蟹不如江蟹，江蟹不如溪蟹，溪蟹不如湖蟹。長江中下游的湖泊中，盛產清水大閘蟹，其中以陽澄湖和太湖的大閘蟹最為著名。每年農曆九月和十月，正是大閘蟹當造的季節，「秋盡江南蟹正肥」，此時源源上市的大閘蟹，個體肥大，膏油豐滿，蟹肉鮮美，正如俗語有云：「吃了大閘蟹，百菜無滋味」。

　　養殖和銷售大閘蟹的蟹農和商家，除了出售整隻活生生的大閘蟹之外，都會把部份體形較小的大閘蟹蒸熟，拆出蟹肉和蟹黃混合起來以盒裝出售，這便是蟹粉。蟹粉味道極為鮮甜，配搭麵食、菇類、蔬菜、豆腐，成為一道道嫩滑鮮美的菜餚，例如蟹粉豆腐、蟹粉獅子頭、炒蟹粉、蟹粉菜心等，而其中蟹粉小籠包和蟹粉湯包更是馳名中外，把蟹粉加在豬肉餡料中，再用老雞和豬皮熬湯做成凍粒，做成灌湯的包子餡。小籠包和湯包的餡和皮基本上相同，只是湯包要比小籠包大，小籠包裏的是汁，湯包裏的是湯。

　　蟹粉配上豆腐最是絕配，膏黃似橙，肉白如雪，鮮甜甘腴，是最受歡迎的江南蟹粉菜餚。蟹粉豆腐最常見的菜式有兩種，一種是做蟹粉豆腐羹，一種是蟹粉扒豆腐。現在市場上有盒裝蟹粉出售，做蟹粉菜式就容易多了。

材料

嫩豆腐 1 塊（約 250 克）　　清雞湯 750 毫升

青豆 20 克　　　　　　　　鹽 1 茶匙

薑米 30 克　　　　　　　　藕粉 2 湯匙

蟹粉 6 湯匙

做法

1. 豆腐切成 0.5 厘米的豆腐粒。

2. 把青豆用水灼熟，瀝乾水分。

3. 用中火燒熱 2 湯匙油，爆香薑米，放入蟹粉輕輕炒勻。

4. 把青豆、豆腐粒和清雞湯放入，煮沸後轉小火煮約 3 分鐘，
   加鹽。

5. 用一小碗，放 4 湯匙水，把藕粉拌勻成芡，徐徐倒入湯中勾芡，
   即成。

Serves: 4-6 / **Preparation time:** 10 minutes / **Cooking time:** 10 minutes

Ingredients

1 soft tofu (about 250 g)

20 g peas

30 g chopped ginger

6 tbsp tomalley and crab meat

750 ml chicken broth

1 tsp salt

2 tbsp lotus root starch

Method

1. Dice tofu into 0.5 cm cubes.

2. Blanch and drain peas.

3. Heat 2 tablespoons of oil over medium heat, stir fry ginger until aromatic, add tomalley and crab, and stir gently to mix.

4. Put in peas, tofu and chicken broth, bring to a boil and reduce to low heat to cook for about 3 minutes. Flavor with salt.

5. Mix lotus root starch with 4 tablespoons of water in a small bowl, and slowly stir into the soup to thicken.

翠光登御舟，入裏湖，出斷橋，又至珍珠園，太上命盡買湖中龜魚放生，並宣喚在湖賣買等人。內侍用小彩旗招引，各有支賜。時有賣魚羹人宋五嫂對御自稱：東京人氏，隨駕到此。太上特宣上船起居，念其年老，賜金錢十文、銀錢一百文、絹十匹，仍令後苑供應泛索。」宋五嫂受太上皇御賜，宋嫂魚羹從此遠近馳名。

宋代吳自牧著的《夢粱錄》第十三卷「鋪席」裏記載了當年杭州大街上的店舖，列出一百多家杭城市肆名家有名者，其中便有錢塘門外宋五嫂魚羹，這就是宋五嫂出了名後，從湖面上搬到岸邊開的飯館。吳自牧出生時，已經是宋高宗御賞宋嫂魚羹後的一百年，文中所記載在杭州大街上的宋五嫂魚羹，應該是後人經營的店舖。

●材料

大桂花魚（鱖魚）1 條（約 700 克）

火腿 30 克

冬菇 5 朵

筍 50 克

蛋白 2 個

薑片 5 克

薑汁 2 湯匙

薑絲 30 克

蔥 2 條

紹酒 2 湯匙

鹽 2 茶匙

鎮江醋 2 湯匙

胡椒粉 1/2 茶匙

西湖藕粉 1.5 湯匙

●烹調心得

1. 可選用桂花魚、鯇魚、黃花魚等來做。

2. 蛋白經過一個過濾網流到煮沸的湯中，便可形成幼絲。

3. 這道菜是羹，所以冬菇和筍要切成幼絲，不能切得太粗。

4. 勾芡用的西湖藕粉，也可用馬蹄粉代替。

份量　6-8 人份

準備時間　20 分鐘

烹調時間　1 小時

① ② ③ ④ ⑤ ⑥ ⑦ ⑧ ⑨ ⑩ ⑪ ⑫

● 做法

1. 在桂花魚靠近魚頭的地方切一刀①，再在近尾的地方切一刀②，用刀沿着魚脊骨把肉起出③。把魚翻過來，同樣方法起出另一邊魚肉④。

2. 把魚頭和骨斬件放在鍋中，加入薑片⑤和 750 毫升清水，大火煮沸後，改中火煮半小時。

3. 把魚頭和骨等湯渣棄去不要⑥，魚湯留用。

4. 把魚肉放在蒸碟中，用 1/2 茶匙鹽和薑汁醃好魚肉⑦，魚皮朝向下，隔水蒸 6-7 分鐘取出，倒去魚水。

5. 用筷子拆散魚肉⑧、檢走細魚骨和魚皮。

6. 冬菇浸發透，切成細絲，筍和葱切絲，火腿蒸 3 分鐘後剁茸。

7. 把蛋白用筷子打勻但不要起泡。

8. 大火燒熱 1 湯匙油，放入薑絲、冬菇絲和筍絲同爆炒⑨，灒紹酒炒勻，再加入魚湯⑩、胡椒粉和鹽，煮沸。

9. 把魚肉和火腿絲倒入⑪，熄火。把蛋白倒入過濾網，徐徐順鍋的圓形倒進鍋內⑫，不要攪動，使之形成白色的蛋白絲。

10. 把羹煮沸，加入鎮江醋稍為拌勻。

11. 藕粉加 3 湯匙溫水拌勻，倒入羹中勾芡，令湯羹變稠，盛出。

12. 撒上火腿茸和葱絲，即可上桌。

Ingredients

1 Mandarin fish (about 700 g)

30 g Jinhua ham

5 dried black mushrooms

50 g bamboo shoot

2 egg whites

5 g ginger slices

2 tbsp ginger juice

30 g ginger shredded

2 stalks scallion

2 tbsp Shaoxing wine

2 tsp salt

2 tbsp Zhenjiang vinegar

1/2 tsp ground white pepper

1.5 tbsp Westlake lotus starch

Serves: 6-8

Preparation time: 20 minutes

Cooking time: 1 hour

Method

1. Make a deep cut near the head ① and another near the tail ② , run the knife along the back and fillet the fish ③ . Turn over the fish and fillet the other side ④ .

2. Cut fish head and bones into several pieces and place into a pot with the ginger slices ⑤ and 750 ml of water, bring to a boil over high heat, then reduce to medium heat and boil for 30 minutes.

3. Drain the fish soup and discard the bones ⑥ .

4. Place fish fillets skin facing down on a plate, marinate with 1/2 teaspoon salt and ginger juice ⑦ , and steam for about 6 to 7 minutes. Pour out water in the plate.

5. Pick out the bones and remove the skin, and shred the meat with chopsticks ⑧ .

6. Soak mushrooms until soft and cut into fine shreds. Similarly cut bamboo shoot and scallion into fine shreds. Steam ham for 3 minutes and finely chopped.

7. Beat egg whites but without making a lot of foam.

8. Heat 1 tablespoon of oil in a pot, stir in shredded ginger, mushroom and bamboo shoot ⑨ , sprinkle wine, add fish soup ⑩ , ground white pepper and salt, and bring to a boil.

9. Stir in fish and ham ⑪ , and turn off the heat. Run egg whites slowly through a wire strainer into the soup in a circular motion ⑫ . Do not stir to allow egg whites to form fine shreds.

10. Bring the soup to a boil and stir in Zhenjiang vinegar.

11. Mix lotus root starch with 3 tablespoon of warm water and gradually stir into the soup to thicken.

12. Top with chopped ham and shredded scallion.

Tips:

1. Other fish such as grass carp and corvina can also be used.

2. Running egg whites through a wire strainer into the soup will allow the egg whites to form into fine lines.

3. Both mushrooms and bamboo shoot should be cut into fine shreds.

4. Lotus root starch can be replaced by water chestnut starch.

雞茸雪蛤

Hashima and Chicken Soup

份量 6 人份
準備時間 10 分鐘
烹調時間 15 分鐘

雪蛤、飛龍、熊掌、猴頭菇被譽為黑龍江省四大山珍。雪蛤又名哈士蟆，屬林蛙類，生長於我國東北嚴寒的山林中，是受國家保護的兩棲動物。雪蛤每年深秋之後就會冬眠，為繁殖下一代，母雪蛤會盡量攝取營養，以至體內積藏很多脂肪，卵巢也儲存大量荷爾蒙，準備作為來年春季產卵之用，這時候就是捕捉雪蛤的季節。捉到這種雌性雪蛤之後，宰之並吊至風乾，再在其腹中取輸卵管和卵巢間的膠狀脂肪。這粒膠狀脂肪就是雪蛤膏，又稱哈士蟆油。

雪蛤膏之珍貴，因為它含豐富的營養成份，包括蛋白質、脂肪、磷、硫，及維生素Ａ、Ｂ、Ｃ，含有人體必需的十八種氨基酸，具補腎、治肺虛、防衰老、養顏之功效，古今中醫皆視之為集食、藥、補於一體的珍品。

現在吃雪蛤膏很方便，一般海味店出售的雪蛤，已取出膠狀雪蛤膏，不用買整隻雪蛤加工。有時會買到的雪蛤膏是帶着輸卵管和卵巢的，作為補品而言，輸卵管和卵巢也是可以吃的，而且營養豐富。浸發好的雪蛤膏，可做成湯羹或甜品，例如「雪蛤燉木瓜」、「雪蛤蓮子羹」，以下介紹的就是「雞茸雪蛤」。

●浸發雪蛤膏及去腥味的方法

雪蛤膏（哈士蟆油）的吸水力特強，要用大量清水浸發，雪蛤膏發起時，體積比乾貨脹大幾十倍，所以要用一個較大的容器來浸發。發起的雪蛤膏，像一團團的荔枝肉，雪白中帶有黑色絲狀物體及小黑點，要用鉗子小心挾走，然後換水漂洗一次，再用沸水沖浸片刻撈出。大火燒沸半鍋水，放幾片薑片或薑汁，加少許紹酒，煮沸後加入雪蛤膏，再煮沸後倒清水中撈洗，取出瀝乾水分，放涼後用食物盒分裝好，放在冰箱冷格中保存，隨時解凍即可用。

材料

已浸發的雪蛤 100 克

薑汁 2 湯匙

新鮮雞胸肉 150 克

蛋白 1 個

雞湯 750 毫升

鹽 1 茶匙

胡椒粉 1/2 茶匙

馬蹄粉 20 克

麻油 1 茶匙

熟火腿茸 20 克

做法

1. 把浸發好的雪蛤用水加薑汁焯 3 分鐘，撈出瀝水，備用。

2. 挑去雞胸的皮筋和膜，剁爛成茸，放在大碗中。

3. 把蛋白拌入雞茸中，用筷子打勻，挑出筋膜不要。

4. 煮沸 250 毫升雞湯，趁熱倒入雞茸，同時不斷向同一方向攪動，至雞茸半熟而完全溶入雞湯中。

5. 把其餘 500 毫升雞湯煮沸，加入煨發好的雪蛤煮 3 分鐘，加鹽和胡椒粉調味，用馬蹄粉加水成芡倒入，煮至湯稠。

6. 改小火，把雞茸湯徐徐倒下，攪勻立即熄火，加入麻油，盛大湯碗，灑上火腿茸，即成。

●烹調心得

1. 加了蛋白的雞茸，用筷子打勻，筋膜可能會纏住筷子，要立即挑出，這樣雞茸就會更加嫩滑。

2. 煮雞茸的火候不能大，微沸即可。

Serves: 6 / **Preparation time:** 10 minutes / **Cooking time:** 15 minutes

Ingredients

100 g pre-soaked hashima

2 tbsp ginger juice

150 g fresh chicken breast

1 egg white

750 ml chicken broth

1 tsp salt

1/2 tsp ground white pepper

20 g water chestnut starch

1 tsp sesame oil

20 g chopped ham

Method

1. Blanch hashima for 3 minutes with ginger juice added. Drain.

2. Remove any tendons and membranes from the chicken breast and finely chop chicken into a fine paste. Put into a large bowl.

3. Stir in egg white with chopsticks, mix well and remove any remaining tendon or membrane attached to the chopsticks.

4. Boil 250 ml of chicken broth, slowly pour into the bowl containing the chicken paste and stir in a single direction until the chicken paste is fully blended with the broth to become a chicken paste broth.

5. Boil the remaining chicken broth, add hashima and cook for about 3 minutes. Season with salt and pepper. Make a wet starch with water chestnut starch by adding 4 tablespoons of water and stir gradually into the soup to thicken.

6. Reduce to low heat, add chicken paste broth gradually, stir gently to blend and turn off the heat. Add sesame oil and put into a soup tureen. Top with chopped ham before serving.

Tips:

1. After egg whites is stirred with chopsticks into the chicken paste, membranes and ligaments from the paste will adhere to chopsticks. They should be removed immediately.

2. Low heat should be used when adding chicken paste broth to the hashima soup.

玉 糝 羹

Rice Soup with White Turnip

　　宋代林洪撰寫的《山家清供》，有一段「玉糝羹」，寫的是才子蘇東坡的食事，原文曰：「東坡一夕與子由飲，酣甚，槌蘆菔爛煮，不用他料，只研白米為糝食之。忽停箸撫几曰：若非天竺酥酏，人間決無此味」。意思是蘇東坡有次與蘇轍（字子由）飲酒，飲到興致甚高之時，把蘆菔（蘿蔔）煮爛，不放其他材料，只是把米研碎做糝吃。席間蘇東坡忽然放下筷子，拍打桌子，說：「除非有天竺酥酏，人間再絕無此美味！」蘇東坡為此羹命名為「玉糝羹」，並作詩稱頌：「香似龍涎仍釅白，味如牛乳更全清。莫將南海金齏膾，輕比東坡玉糝羹。」（見《蘇軾集注》）。

　　糝（普通話音為「身」Shen，粵音為「審」Saam），古代以穀物煮的湯叫做糝。在這道玉糝羹中，是指米糝，俗稱米踤子，即磨成粗粉粒狀的米。現代中菜技巧中，也有所謂「打糝」或「製糝」，即打肉茸，例如四川和一些省份，俗稱茸為「糝」，稱肉茸為「肉糝」，蝦茸為「蝦糝」，魚茸為「魚糝」。

●材料

白蘿蔔 1 個（約 600 克）

白米 50 克

鹽 1 茶匙

●做法

1. 白米洗淨，用攪拌機磨成粗粒。

2. 在鍋內煮沸 1.5 公升水，倒入粗米粒，大火沸煮 45 分鐘成米湯。

3. 白蘿蔔削皮後磨成茸，放入米湯內煮 5 分鐘，加鹽調味即成。

Serves: 4 / Preparation time: 10 minutes / Cooking time: 1 hour

---

Ingredients

600 g white turnip

50 g white rice

1 tsp salt

Method

1. Clean rice and crush into thick pellets with a blender.

2. Bring to a boil 1500 ml of water, add rice, and boil over high heat for 45 minutes.

3. Peel and grate turnip, add to the rice soup and boil for 5 minutes. Add salt to flavor.

份量　4 人份

準備時間　10 分鐘

烹調時間　1 小時

「玉井飯」是一道很古老的杭州飯食，與「蟹釀橙」同樣出自南宋林洪編者的《山家清供》，此書共兩卷，稱得上是我國最早具養生概念的農家菜譜。

《山家清供》記載了一段「玉井飯」的故事，話說林洪認識一位德高望重的長者章鑒（雪齋）先生，雖年事已高，但還是喜歡在家以食物招待來客，不過他大多數的食物都不在市集購買，因為怕打擾到別人。一天，林洪去拜訪老先生，當時剛巧有蝗蟲之災，但卻沒有蝗蟲飛入老先生的家，老先生大為欣慰。為慶祝這件值得感恩的事，老先生留林洪晚酌數杯，吩咐下人以新鮮的蓮藕和蓮子來煮「玉井飯」。

《山家清供》文中記載了做法，原文如下：「削嫩白藕作塊，採新蓮子去皮、芯，候飯少沸，投之，如罨飯法。」蓮藕和蓮子清甜柔潤，清熱消滯，對身體有益，「玉子飯」不失為養生膳食的好主意！

份量　4 人份

準備時間　20 分鐘

烹調時間　一般煮飯時間

玉 井 飯

Rice with Lotus Root and Seeds

●**材料**

蓮藕 250 克

新鮮蓮子 100 克

白米 320 克

●**做法**

1. 蓮藕切去藕節，削皮，洗淨，再切成約 1 厘米立方粒。

2. 把蓮子的皮剝掉，再把芯挑出不要。蓮子洗淨後備用。

3. 白米洗淨，下鍋，放入適量水，待米水煮沸，放下蓮藕粒和蓮子，
   與米拌勻，煮至飯熟。

●**烹調心得**

如果買不到新鮮蓮子，也可以用乾蓮子或冰鮮蓮子代替。乾蓮子要先用清水泡 1 小時，

再挑出蓮子芯不要。冰鮮蓮子一般已經去芯，下鍋前用清水泡 5 分鐘。

Serves: 4 / Preparation time: 20 minutes / Cooking time: normal rice cooking time

Ingredients

250 g lotus root

100 g fresh lotus seeds

320 g rice

Method

1. Cut off the knots of the lotus root, peel and clean, and cut into 1 cm cubes.

2. Peel and remove the bitter germ of the lotus seeds. Rinse and drain seeds.

3. Cook rice by the normal method, and add lotus roots and seeds when water has come to a boil. Continue to cook until rice is done.

Tips:

Dried or chilled lotus seeds can be used if fresh lotus seeds are not available. Dried seeds should be soaked for 1 hour first and then pick out the bitter germ. Chilled lotus seeds usually have their bitter germs removed but should be soaked in fresh water for 5 minutes before use.

雪菜黃魚煨麵

Noodles with Corvina Fish and Potherb Mustard

份量 2 人份
準備時間 15 分鐘
烹調時間 30 分鐘

　　「上有天堂，下有蘇杭」，如此美譽，當然一定離不開美食。杭州的小食多彩多姿，琳瑯滿目，有小籠包、燒賣、餃子、饅頭、湯丸、油條、餛飩、酥油餅、西湖藕粉和各式麵食，而杭州的包子和麵食，更是全國聞名。做包子和麵食的小麥，本來是北方人的主要糧食，由於宋室南遷，使江南地區的飲食習慣起了很大的變化，而包子和麵食，成為了杭州庶民百姓最喜愛的食物。

　　北方人吃麵條，吃的是麥的香味和麵的筋道，麵的本身就是主角，所以傳統的北方人，家家會做麵條，不像南方人買現成的麵就算了。而杭州人吃麵條，講究的是配麵的湯和配麵的材料。杭州是魚米之鄉，風調雨順，鮮美的食材隨手可得，杭州人得天獨厚，性格是自得其樂，優哉悠哉。他們會一絲不苟地炒一小盤菜，或者熬一小鍋湯，繁複多樣，就是為了吃一碗心滿意足的好麵，也只有杭州人能有如此能耐，這種生活品味的神韻，如非杭州人是無法心領會的。

●材料

小黃魚 450 克

鹽 1 茶匙

薑汁 1 湯匙

胡椒粉 1/8 茶匙

薑片 5 片

酒 1 茶匙

沸水 500 毫升

雪裡蕻 100 克

糖 1 茶匙

上海幼麵 / 掛麵 300 克

筍絲 50 克

做法

1. 將黃魚起肉，魚頭及魚骨留用。用鑷子把魚肉的小骨拔去，洗淨瀝水，灑上 1/2 茶匙鹽，醃約 15 分鐘，用廚紙吸乾鹽和水分，抹上薑汁及胡椒粉。

2. 在鍋裏燒熱 2 湯匙油，放進魚肉，先煎魚皮至金黃，再煎另一面，取出備用。

3. 放入魚頭、魚骨和薑片，煎香後潷酒，加入沸水，大火煮至魚湯轉白色，把魚湯過濾去渣，湯留用。

4. 雪裡蕻切去頭部及前段硬莖不要，沖洗後切成碎粒。燒熱鑊，不加油，白鑊焙乾雪裡蕻，加糖炒勻，取出備用。

5. 上海幼麵用水焓至軟身僅熟，瀝水備用。

6. 煮沸魚湯，放入雪裡蕻和筍絲，沸煮 5 分鐘。

7. 把焓好的麵條放入湯中，放入魚肉。用小火煨煮約 5 分鐘，可加鹽調味，即成。

Serves: 2 / Preparation time: 15 minutes / Cooking time: 30 minutes

Ingredients

450 g Corvina fish

1 tsp salt

1 tbsp ginger juice

1/8 tsp ground white pepper

5 ginger slices

1 tsp wine

500 ml boiling water

100 g preserved potherb mustard

1 tsp sugar

300 g Shanghai noodles

50 g shredded bamboo shoots

Method

1. Fillet fish and save the head and bones for later use. Use a pair of tweezers to remove bones from the fillet, then rinse and drain. Marinate fillet with 1/2 teaspoon of salt for 15 minutes. Dry with kitchen towels and mix with ginger juice and ground white pepper.

2. Heat 2 tablespoons of oil in a pot, pan fry fillets with skin side down until golden brown, flip over the fillets and pan fry until fully cooked. Remove fillets.

3. With the oil remaining in the pot, pan fry ginger, fish head and bones, add wine and boiling water and boil over high heat until soup turns a white color. Filter soup and save for later.

4. Cut off and discard the root portion of the potherb mustard greens then rinse and chop. Roast potherb mustard in a dry wok to remove most of the moisture and stir in sugar.

5. Cook noodles and drain.

6. Boil fish soup, add potherb mustard and bamboo shoots, and boil for 5 minutes.

7. Add noodles to the soup top with fish fillets, and simmer for 5 minutes over low heat. Flavor with additional salt if necessary.

不少杭州的方言，在名詞裏加個「兒」字，是由北方傳過來的，本身並不是南方的語言習俗，傳統的北京話，同樣也帶不少的「兒」字。杭州話中帶「兒」字的，例如棒兒糖、顆兒糖、芡兒粉、貓兒朵、筒兒骨、踏兒哥、柯兒紙、豆兒鬼、門兒布等等，以及以下介紹的片兒川。

片兒川是杭州百年老店奎元館的風味名點，以鮮筍片、豬肉片和雪裡蕻（雪菜）做湯麵的餸料（澆頭）煮麵而成，因傳統上這三種材料都是用水焯熟，即汆熟，而「汆」字的音與「川」字同音，所以杭州人稱這種傳統湯麵為片兒川。

片兒川是一道家常麵點，街道小吃，也是杭州宴客菜中，最常見的單尾主食，吃完豐富多姿的杭幫菜，最後來一碗熱呼呼的片兒川麵，為宴會劃上韻味十足的完美句號。

片兒川麵

Pianerchuan Noodles

材料

豬里脊 60 克	筍 60 克
生抽 1 茶匙	乾上海幼麵 150 克
生粉 1/2 茶匙	清雞湯 600 毫升
雪裡蕻 60 克	鹽適量（調味用）

份量 4 人份
準備時間 10 分鐘
烹調時間 15 分鐘

做法

1. 豬肉洗淨切片，先用生抽拌勻，再拌入生粉。
2. 雪裡蕻切去菜頭，洗淨，擠乾水後再切成絲，筍切筍片，備用。
3. 在鍋裏燒滾 2 公升水，把麵煮熟，取出，過冷河。
4. 在鍋裏煮沸雞湯，下豬肉、雪裡蕻和筍片，煮沸，用適量的鹽調味。
5. 放入煮好的麵條，再煮沸即成。

Serves: 4 / Preparation time: 10 minutes / Cooking time: 15 minutes

Ingredients

60 g pork tenderloin

1 tsp light soy sauce

1/2 tsp corn starch

60 g pickled potherb mustard

60 g bamboo shoot

150 g dried Shanghai noodles

600 ml chicken broth

salt (flavoring)

Method

1. Slice pork, mix with 1 teaspoon of soy sauce and then stir in corn starch.
2. Rinse potherb mustard after removing the head, squeeze out excess water and cut into shreds. Cut bamboo shoot into slices.
3. Cook noodles in about 2 liters of boiling water, drain, and rinse with cold water.
4. Bring to a boil chicken broth, add pork, potherb mustard and bamboo shoots, re-boil and flavor with salt.
5. Put in noodles and re-boil.

桂花鮮栗羹

Sweet Chestnut Soup with Osmanthus Flower

份量　4 人份
準備時間　10 分鐘
烹調時間　10 分鐘

金桂飄香栗子來，每年桂花開的時候，栗子也上市了。栗子肉粉糯甜美，更有滋補功能，素有「千果之王」的稱譽，自古醫者文人對吃栗子的好處推崇備至，唐代醫藥名家孫思邈說栗子是「腎之果也，腎病宜食之」。

「桂花鮮栗羹」是杭州著名的甜品小吃，據說來自一個美麗的傳說。話說唐朝時期的一個中秋夜，寂寞的嫦娥在月亮的廣寒宮中凝望人間，為西湖美景所吸引，於是舒展廣袖翩翩起舞，吳剛則在旁手擊桂樹為她伴奏，震落滿天桂籽灑到人間。正在這時候，杭州靈隱寺的德明和尚正在煮栗子粥，來自月宮的桂籽飄落在粥中，其香無比，眾僧品嘗後大讚。第二天德明和尚便將地上的桂籽收集起來，種在西湖附近的山坡上。桂樹生長得很快，在八月時更開滿了金黃色、白色和紅色的小花，色彩爛漫，香氣襲人，這就是西湖美景中的金桂、銀桂、丹桂。而「桂花鮮栗羹」也隨着神話代代相傳，成了杭州美食。

材料

栗子肉 200 克
白糖 4 湯匙
西湖藕粉 20 克
桂花糖 20 克

做法

1. 去殼去衣的鮮栗子肉切成 4 塊，小火用水煮 5 分鐘至熟，撈出。
2. 煮沸 500 毫升清水，加入糖和栗子肉同煮，潷去浮沫，煮 3 分鐘。
3. 用 50 毫升溫水，加入藕粉調和，再把藕粉水慢慢注入煮沸的栗子湯中，邊沖邊攪，至沒有粉粒，調成半透明羹狀，分盛在四個碗中。
4. 分別淋上桂花糖，即成

Serves: 4 / Preparation time: 10 minutes / Cooking time: 10 minutes

Ingredients

200 g peeled chestnuts
4 tbsp sugar
20 g Westlake lotus starch
20 g osmanthus sugar

Method

1. Cut peeled chestnuts into 4 pieces and boil over low heat for about 5 minutes. Drain.
2. Add sugar and chestnuts to 500 ml of boiling water and boil for 3 minutes and skim any foam from the surface.
3. Mix lotus starch with 50 ml of warm water thoroughly and stir into the boiling chestnut soup gradually until all particles are dissolved. Put into individual bowls.
4. Top with osmanthus sugar before serving.

度 量 衡 換 算 表

　　市面上的食譜書，包括我們陳家廚坊系列，食譜中的計量單位，都是採用公制，即重量以克來表示，長度以厘米 cm 來表示，而容量單位以毫升 ml 來表示。世界上大多數國家都採用公制，但亦有少數國家如美國，至今仍使用英制（安士、磅、英吋、英呎）。

　　香港和澳門，一般街市仍沿用司馬秤（斤、兩），在香港超市則有時用公制，有時會用美制，所以香港是世界上計量單位最混亂的城市，很容易會產生誤會。與香港關係緊密的中國大陸，他們的大超市有採用公制，但一般市民用的是市制斤兩，這個斤與兩，實際重量又與香港人用的司馬秤不同。

　　鑒於換算之不方便，曾有讀者要求我們在食譜中寫上公制及司馬秤兩種單位，但由於編輯排版困難，實在難以做到。考慮到實際情況的需要，我們覺得有必要把度量衡的換算，以圖表方式來說清楚。

重量換算速查表 （公制換其他重量單位）

克	司馬兩	司馬斤	安士	磅	市斤
1	0.027	0.002	0.035	0.002	0.002
2	0.053	0.003	0.071	0.004	0.004
3	0.080	0.005	0.106	0.007	0.006
4	0.107	0.007	0.141	0.009	0.008
5	0.133	0.008	0.176	0.011	0.010
10	0.267	0.017	0.353	0.022	0.020
15	0.400	0.025	0.529	0.033	0.030
20	0.533	0.033	0.705	0.044	0.040
25	0.667	0.042	0.882	0.055	0.050
30	0.800	0.050	1.058	0.066	0.060
40	1.067	0.067	1.411	0.088	0.080
50	1.334	0.084	1.764	0.111	0.100
60	1.600	0.100	2.116	0.133	0.120
70	1.867	0.117	2.469	0.155	0.140
80	2.134	0.134	2.822	0.177	0.160
90	2.400	0.150	3.174	0.199	0.180
100	2.67	0.17	3.53	0.22	0.20
150	4.00	0.25	5.29	0.33	0.30
200	5.33	0.33	7.05	0.44	0.40
250	6.67	0.42	8.82	0.55	0.50
300	8.00	0.50	10.58	0.66	0.60
350	9.33	0.58	12.34	0.77	0.70
400	10.67	0.67	14.11	0.88	0.80
450	12.00	0.75	15.87	0.99	0.90
500	13.34	0.84	17.64	1.11	1.00
600	16.00	1.00	21.16	1.33	1.20
700	18.67	1.17	24.69	1.55	1.40
800	21.34	1.34	28.22	1.77	1.60
900	24.00	1.50	31.74	1.99	1.80
1000	26.67	1.67	35.27	2.21	2.00

司馬秤換公制

司馬兩	司馬斤	克
1		37.5
2		75
3		112.5
4	0.25	150
5		187.5
6		225
7		262.5
8	0.5	300
9		337.5
10		375
11		412.5
12	0.75	450
13		487.5
14		525
15		562.5
16	1	600
24	1.5	900
32	2	1200
40	2.5	1500
48	3	1800
56	3.5	2100
64	4	2400
80	5	3000

英制換公制

安士	磅	克
1		28.5
2		57
3		85
4	0.25	113.5
5		142
6		170
7		199
8	0.5	227
9		255
10		284
11		312
12	0.75	340.5
13		369
14		397
15		426
16	1	454
24	1.5	681
32	2	908
40	2.5	1135
48	3	1362
56	3.5	1589
64	4	1816
80	5	2270

容量

量杯	公制（毫升）	美制（液體安士）
1/4 杯	60 ml	2 fl. oz.
1/2 杯	125 ml	4 fl. oz.
1 杯	250 ml	8 fl. oz.
1 1/2 杯	375 ml	12 fl. oz.
2 杯	500 ml	16 fl. oz.
4 杯	1000 ml /1 公升	32 fl. oz.

量匙	公制（毫升）
1/8 茶匙	0.5 ml
1/4 茶匙	1 ml
1/2 茶匙	2 ml
3/4 茶匙	4 ml
1 茶匙	5 ml
1 湯匙	15 ml

鳴謝

深圳西湖春天飲食集團　　冠珍（興記）醬園　　　三陽泰　　　Inhesion (Asia) Ltd.

張曉光先生　　　　　　　　傅月良先生

黃可尚先生　　　　　　　　張繼鋼先生

鄧子怡女士

參考古代文獻資料

司馬遷　《史記》　漢朝　　　　　　　林洪　《山家清供》　南宋

賈思勰　《齊民要術》　北魏　　　　　浦江吳氏　《吳氏中饋錄》　南宋

孟元老　《東京夢華錄》　北宋　　　　李時珍　《本草綱目》　明代

蘇東坡　《蘇軾集注》　南宋　　　　　袁枚　《隨園食單》　清代

吳自牧　《夢粱錄》　南宋

參考現代著作資料

作者林惠祥　《中國民族史》　1937年　上海書店

作者楊曉東　《燦爛的吳地魚稻文化》1993年　北京：當代中國出版社

作者王仁湘　《飲食之旅》1993年　北京人民出版社

作者茅天堯《品味紹興》2005年浙江科學技術出版社

張宇光主編　《中華飲食文獻匯編》2007年中國國際廣播出版社

趙榮光主編　《中國飲食文化史》2013年　中國輕工業出版社

胡忠英主編　《杭州南宋菜譜》　2013年　浙江人民出版社

作者介紹

方曉嵐、陳紀臨夫婦為近代著名飲食文化作家陳夢因（特級校對）兒媳，傳承陳家兩代的烹飪知識，對飲食文化作不懈的探討研究，作品內容豐富實用，文筆流麗，深受讀者歡迎，是香港暢銷的食譜書作家，至今已在香港出版了 13 本食譜書，作品遠銷海外及國內市場，更在台灣多次出版。

2016 年方曉嵐、陳紀臨夫婦應英國著名出版商 Phaidon Press 的邀請，用英文撰寫了 *China The Cookbook*，介紹中國 33 個省市自治區的飲食文化和超過 650 個各省地道菜式的食譜，這本書得到國際上很多好評，並為世界各大主要圖書館收藏，現正翻譯成法文、德文、中文、意大利文、荷蘭文、西班牙文等多國文字；陳氏伉儷並往加拿大、美國、英國及澳洲等國多個大城市進行巡迴推廣、演講及接受傳媒採訪，以香港作家的身份，為中國菜在國際舞台上的發展作出貢獻，為香港人爭光。

如有查詢，請登入：

f 陳家廚坊讀者會

或電郵至：
chanskitchen@yahoo.com

迴味・杭州菜 Hangzhou Cuisine

作者 Author
陳家廚坊 Chan's Kitchen
方曉嵐・陳紀臨 Diora Fong • Keilum Chan

策劃／編輯 Project Editor
Catherine Tam

攝影 Photographer
傅光
葉冲
Imagine Union

美術設計 Design
Charlotte Chau

出版者 Publisher
萬里機構出版有限公司 Wan Li Book Co Ltd.
香港鰂魚涌英皇道 1065 號 Room 1305, Eastern Centre, 1065 King's Road,
東達中心 1305 室 Quarry Bay, Hong Kong
電話 Tel 2564 7511
傳真 Fax 2565 5539
電郵 Email info@wanlibk.com
網址 Web Site http://www.wanlibk.com
http//www.facebook.com/wanlibk

發行者 Distributor
香港聯合書刊物流有限公司 SUP Publishing Logistics (HK) Ltd.
香港新界大埔汀麗路 36 號 3/F., C&C Building, 36 Ting Lai Road,
中華商務印刷大廈 3 字樓 Tai Po, N.T., Hong Kong
電話 Tel 2150 2100
傳真 Fax 2407 3062
電郵 Email info@suplogistics.com.hk

承印者 Printer
中華商務彩色印刷有限公司 C & C Offset Printing Co., Ltd.

出版日期 Publishing Date
二零一七年七月第一次印刷 First print in July 2017